THE APOCALYPSE FACTORY

ALSO BY STEVE OLSON

MAPPING HUMAN HISTORY:
GENES, RACE, AND OUR COMMON ORIGINS

COUNT DOWN:
SIX KIDS VIE FOR GLORY AT THE WORLD'S
TOUGHEST MATH COMPETITION

ANARCHY EVOLUTION:
FAITH, SCIENCE, AND BAD RELIGION IN
A WORLD WITHOUT GOD
(coauthored with Greg Graffin)

ERUPTION:
THE UNTOLD STORY OF MOUNT ST. HELENS

THE APOCALYPSE FACTORY

PLUTONIUM AND THE MAKING OF THE ATOMIC AGE

STEVE OLSON

W. W. NORTON & COMPANY

Independent Publishers Since 1923

For information about permission to reproduce selections from this book, write to Permissions, W. W. Norton & Company, Inc., 500 Fifth Avenue, New York, NY 10110

For information about special discounts for bulk purchases, please contact W. W. Norton Special Sales at specialsales@wwnorton.com or 800-233-4830

Manufacturing by Lake Book Manufacturing
Book design by Lovedog Studio
Production manager: Julia Druskin

Library of Congress Cataloging-in-Publication Data

Names: Olson, Steve, 1956– author.
Title: The apocalypse factory : plutonium and the making of the atomic age / Steve Olson.
Other titles: Plutonium and the making of the atomic age
Description: First edition. | New York : W. W. Norton & Company, [2020] | Includes bibliographical references and index.
Identifiers: LCCN 2020008293 | ISBN 9780393634976 (hardcover) | ISBN 9780393634983 (epub)
Subjects: LCSH: Atomic bomb—United States—History—20th century. | Plutonium industry—Washington (State)—Hanford—History—20th century. | Hanford Engineer Works—History. | Nuclear weapons—United States—History.
Classification: LCC QC773.A1 O47 2020 | DDC 623.4/51190973—dc23
LC record available at https://lccn.loc.gov/2020008293

W. W. Norton & Company, Inc., 500 Fifth Avenue, New York, N.Y. 10110
www.wwnorton.com

W. W. Norton & Company Ltd., 15 Carlisle Street, London W1D 3BS

1 2 3 4 5 6 7 8 9 0

For Lynn, again,
and in memory of John Hersey

The Alfred P. Sloan Foundation provided support for the
writing of this book through its Public Understanding of
Science, Technology, and Economics program.

Apocalypse: A prophetic revelation of future events.

. . . Have you ever held plutonium
in your hand? Someone once gave me a piece shaped and
   nickel-plated
so alpha particles couldn't reach the skin. It was the temperature,
   you see,
the element producing heat to keep itself warm—not for ten
or a hundred years, but thousands of years. This is the energy
   contained
in Hanford's fuel. I think of that place as a song not properly sung.
A romantic song. And not one person in a hundred knows the tune.

—Kathleen Flenniken,
"A Great Physicist Recalls the Manhattan Project"

CONTENTS

THE APOCALYPSE FACTORY

PROLOGUE

AS SOON AS FRANKLIN MATTHIAS FLEW OVER THE HORSE HEAVEN
Hills, he knew he'd found what he was looking for. Spread out below
him was a barren, wind-swept plain. The Columbia River, running
cold and deep, separated the plain from higher ground to the north
and east. To the west the top of Rattlesnake Mountain was covered
by snow, but the rest of the land, tinted gray-green by sagebrush, was
snow-free, even on this first day of winter.

Matthias, a 34-year-old colonel in the US Army Corps of Engineers, went through the checklist he'd gotten from the engineers at
DuPont. The Columbia could provide plenty of clear, cold water
for the reactors. A row of spindly metal towers carried high-voltage
lines from Grand Coulee Dam, which had come online just the year
before. Not far from the towers, a spur line from the Milwaukee
Road emerged from the gap where the Columbia cut through the
Saddle Mountains. DuPont could haul equipment, construction
materials, and chemicals down the rail line to the site.

The plain was at least twice as large as the 12-mile by 16-mile
expanse that the engineers had demanded, and no large towns were
nearby. If one of the reactors blew up, relatively few people would
be killed. And on that December 22, 1942, with the sun low on the
horizon and silvered by clouds, the land looked sere and forlorn. A
few scraggly towns—Richland, Hanford, and White Bluffs, according to the map—interrupted the treeless expanse, along with some
bedraggled farms. But the number of people who would have to

move couldn't be more than a thousand or two. "This is it," Matthias thought. "There's nothing like it in the country."

Matthias was looking for a place to build a facility that he had been told could end World War II. Four years earlier, two chemists working in a laboratory a few miles away from Hitler's Berlin headquarters, with the invaluable help of an Austrian physicist living in exile in Sweden, had announced that atoms of uranium could split and release immense quantities of energy. Alarmed that Nazi Germany would use the discovery to build atomic bombs, the United States had launched a crash program, dubbed the Manhattan Project, to build them first. Even as Matthias was flying over the towns of Richland, Hanford, and White Bluffs, workers were clearing a vast tract of land in eastern Tennessee, near a sharp rise of land known as Black Oak Ridge, where a massive factory to create bomb-making materials would rise over the next two years. On the flanks of an extinct volcano in New Mexico, other workers were converting a boys' school in the tiny hamlet of Los Alamos into a top-secret laboratory, where many of America's leading scientists would soon congregate to assemble the raw materials of the atomic age into weapons of unprecedented power.

Los Alamos and, to a lesser extent, Oak Ridge have gotten most of the attention in histories of the Manhattan Project. That's understandable. The Oak Ridge facility produced the material in the first atomic bomb used in warfare, the one dropped on Hiroshima, Japan, on August 6, 1945. The scientists at Los Alamos accomplished in a few short years work that would normally have taken much longer.

But in this book I argue that the Hanford nuclear reservation in south-central Washington State is the single most important site of the nuclear age. Hanford produced the material in the first atomic bomb ever exploded, near Alamogordo, New Mexico, on July 16, 1945. The first full-scale nuclear reactor was built at Hanford, and all subsequent reactors have used ideas and technologies developed there. The last atomic bomb used in warfare—the one dropped on

Nagasaki, Japan, on August 9, 1945—contained nuclear material manufactured at Hanford.

The Hiroshima bomb was a technological one-off. No subsequent bombs used that design, except for a few artillery weapons that were soon discarded. At the core of every nuclear weapon in the world today is a small pit of radioactive material made either at Hanford or at a comparable facility elsewhere in the world.

Hanford and its successor facilities have given us, for the first time in history, the ability to destroy ourselves and everything we have ever created. Understanding what happened at the site that Colonel Matthias chose on that solstice day in December 1942 may give us a way to avoid that fate.

IF MATTHIAS HAD LOOKED through the window of his reconnaissance plane toward the northeast, he would have seen a distant gray smudge on the horizon. That's the town of Othello, Washington, where I grew up in the 1960s and early 1970s. I had an idyllic, all-American childhood in Othello. In those days of laissez-faire parenting, my friends and I, on weekends and in the summer, were free to do pretty much anything we wanted once we'd finished a few chores. We could ride our bikes to the swimming holes outside town, to the strange rock formations carved into the desert by prehistoric floods, sometimes all the way to the Tri-Cities, 60 miles to the south, on the banks of the Columbia River. We congregated at Othello's many parks to play baseball, at the drugstore to read comics, and at the one-room library adjacent to city hall. In rosy hindsight, I remember Othello as an isolated, self-contained paradise where we were free to make our own mistakes and enjoy our own triumphs.

If I hadn't lived in Othello, I could have written almost exactly the same book as you hold in your hands. Then again, if I hadn't lived in Othello, this book might never have been written. Hanford and Nagasaki have always been the neglected stepchildren of World War II. Some people might have heard of Hanford as the "most contami-

The first time Colonel Franklin Matthias saw the barren stretch of land extending from Richland to White Bluffs in south-central Washington State, he knew it would be perfect for the plutonium production facilities that came to be known as Hanford. The distance from Richland to White Bluffs is about 30 miles. *Map courtesy of Matt Stevenson; drawing courtesy of Sarah Olson.*

nated nuclear site in the Western Hemisphere," or as the location of a multi-billion-dollar Department of Energy cleanup, but they probably haven't heard much else. In Japan, residents of Nagasaki often wonder why Hiroshima gets so much more attention than the city on which the second bomb was dropped. Throughout the world, people tend to speak of "the bomb," rather than "the bombs," as if the two quite different bombs dropped on Japan can all be captured in a single abstract image.

Hanford was born in secrecy, operated in secrecy, and remains obscure today. Most histories of the Manhattan Project have focused on the exotic physicists who worked largely in New Mexico, not on the chemists, metallurgists, and engineers elsewhere who were just as vital to the project's success. As Richard Rhodes said of his magisterial four-volume history of the atomic age, "I treated plutonium production to some extent as a black box, inadvertently contributing to the myth that the atomic bomb was the work of 30 theoretical physicists at Los Alamos." Yet the successful operation of the facilities built at Hanford changed the course of history. World War II would not have ended the way it did without the bomb-making material from Hanford's reactors. The Cold War would not have been fought the way it was if Hanford and the corresponding facility in the Soviet Union had not been churning out their deadly product—enough to end human civilization many times over. The ongoing cleanup of the site has provided the world with a precautionary tale of nuclear overreach.

The history of the atomic age looks radically different from the perspective of Hanford. The United States probably would not have undertaken an atomic bomb project during World War II if Hanford could not have been built. Without Hanford, the US military certainly would not have had multiple atomic weapons with which to bomb Japan and threaten the Soviet Union after the war. The history of Hanford explains why the world's first nuclear reactor was built when and where it was. It accounts for the decision to bomb Nagasaki just three days after Hiroshima, before the Japanese government had time to react to the new era of warfare that had begun. It recalibrates the moral calculus of the Manhattan Project. When asked after the war about the famous statement made by Robert Oppenheimer, the head of the Los Alamos laboratory, that physicists had "known sin," the lead engineer at DuPont, the company that built Hanford, said, "My God, if everybody that has made an important contribution to the Hanford project was a sinner, then it would take several Rose bowls to hold them all."

Of the three sites in the Manhattan Project National Historical Park established in 2015—Oak Ridge, Los Alamos, and Hanford— the latter is where the physical, the personal, and the political meet most starkly. Hanford represents one of humanity's greatest intellectual achievements; it also embodies a moral blindness that could destroy us all. People have begun to realize, after decades of warnings, that climate change poses a severe threat to our species. Yet they blithely overlook the fact that human civilization could end in

Seen from the top of Rattlesnake Mountain, the land selected for the Hanford nuclear reservation lies within a broad bend of the Columbia River. *Courtesy of the US Department of Energy.*

a few hours if the leaders of the nuclear states were to unleash the force that Hanford has placed in their hands.

Hanford is where people confronted for the first time all the dilemmas of power, pollution, destruction, and sustainability associated with nuclear energy. The laws of physics enable us to create substances that can release immense quantities of energy. This energy can power electrical grids, cure cancer or cause it, and destroy cities. The substances produced in nuclear reactors will remain dangerous for hundreds of thousands of years, longer than we have existed as a species on this planet. Are we mature, reasonable, and wise enough to be entrusted with such knowledge?

What Matthias had been told about his 1942 mission to Washington State was right. The facilities built at Hanford did end World War II. They also opened the door to a new world, one with which we still have not come to terms.

PART 1

THE ROAD
TO HANFORD

"I didn't think, 'My God, we've changed the history
of the world.' "
—Glenn Seaborg

Chapter 1

BEGINNINGS

THE ROAD TO HANFORD BEGAN IN A COLLEGE CLASSROOM IN SOUTHERN California. There, in early 1932, on the recently opened campus of the University of California, Los Angeles, a chemistry student named Glenn Seaborg heard about a momentous scientific development—the discovery of the neutron. Over the next 13 years, the combination of that student and that discovery would lead to entirely new radio-active substances, to entirely new elements, and to atomic bombs.

Toward the end of his long and productive career, *The Guinness Book of World Records* recognized Seaborg as having the longest biography of anyone in *Who's Who in America*. Yet he came from a small and isolated town, was not much interested in science until his last two years in high school, and for a long time questioned his ability to do great things. Compared with European scientists, who often came from privileged families, Seaborg was typical of mid-20th-century US scientists, who tended to be from rural and middle-class families—perhaps, according to one commentator, because "pathways to success in big business or the established professions are relatively unavailable, but social mobility via a scientific educa-tion is possible." As his future colleague Emilio Segrè, whose father had owned a factory near Rome, once exclaimed: "Seaborg's brother is a truck driver!"

In 1922, when he was 10 years old, Seaborg's parents sold their house in Ishpeming, Michigan—a small town about a dozen miles west of Marquette—and bought one-way tickets to California.

They settled into a new subdivision in what is now the city of South Gate near Los Angeles. They carried water in a bucket to their partly finished home, raised vegetables in a backyard garden, and cooked eggs from their chickens. "We weren't poor so much as we just didn't have money—the same as almost everyone in the neighborhood," Seaborg later recalled.

He went to high school two miles away, in an already diverse Watts. Seaborg's mother wanted him to be a bookkeeper. He thought he might be interested in literature. But he knew he wanted to go to college, and a school counselor told him before his junior year that he needed to take at least one laboratory class to qualify. He signed up for chemistry.

His teacher was a man named Dwight Logan Reid, who "taught chemistry with the charisma and enthusiasm of an old-fashioned preacher," Seaborg wrote. Science immediately appealed to Seaborg. It had an internal logic. "You could learn certain principles and then make predictions. It all seemed to hang together." By halfway through the year, he had decided to become a chemist.

He had the highest grades of his 45-member high school class and was admitted to the University of California, Los Angeles, on a scholarship. But even though he could commute from home, the incidental expenses were going to be hard to cover. The summer after high school he got a job at the Firestone Rubber and Tire Company, cleaning the caked tar off glassware and preparing chemical compounds. No matter what else happened, Seaborg thought, he could always become an industrial chemist.

His junior year, Seaborg began taking physics classes in addition to the chemistry classes required for his major. It was a remarkable year to be a college science student. In February of 1932, the discovery of the neutron by the British physicist James Chadwick answered a question people had been asking for millennia: What is the world made of?

Since the turn of the 20th century, scientists had known that matter is made of atoms—minuscule particles so small that 20 million of them lined up in a row would just about span the letter *o*. But the constituents of atoms had remained mysterious. Physicists knew that they contain positively charged particles called protons that are clustered together in a central object called the nucleus. This nucleus is surrounded by a swirling mist of much lighter particles known as electrons. The number of protons determines what kind of element an atom is. Hydrogen, at the top of the list of elements, has a single proton. Uranium, at the bottom of the list (as it then existed), has 92.

But something else had to be going on in atomic nuclei. For one thing, the positively charged protons should repel each other, like the matching ends of two bar magnets. What was keeping the nucleus together? Also, physicists had determined that atoms of the same element could have different weights, so nuclei had to contain something besides protons, but what?

The discovery of the neutron solved both problems at once. With their neutral electric charge, neutrons act as a sort of nuclear glue. They stick to protons and to each other to keep the protons in nuclei from blowing atoms apart. Furthermore, atoms with the same number of protons can have different numbers of neutrons, which accounts for the different weights of what are known as isotopes of the same element. It all made sense. Everything fit together.

Scientists were delighted by the discovery of the neutron. It explained how atoms are constructed and why they have the properties they do. But the neutron's discovery also sent science—and incipient scientists like Seaborg—careening down entirely new paths.

AS A UCLA FRESHMAN, Seaborg had no idea what a PhD degree was. In his junior year, he decided he wanted to get one. He applied to the graduate program in chemistry at the University of California, Berkeley, and, despite his "lingering disbelief" that he belonged at such a top-notch university, was accepted. "The whole Berkeley atmosphere was like magic to me," he later wrote. "Berkeley was Wonderland."

When Seaborg began graduate school in the fall of 1934, Berkeley was one of the leading universities for science anywhere in the world. The chemistry department, under the leadership of its dean, Gilbert Lewis, was widely considered the best in the United States. Seaborg soon impressed the gruff, cigar-chomping Lewis with his good sense and industriousness. He later spent many hours at Berkeley preparing solutions and making measurements under Lewis's watchful instruction—critical preparation for the tasks to come.

The physics department at Berkeley was even stronger. On the experimental side, it centered on Ernest Lawrence, the inventor of a device called a cyclotron. Though Lawrence was 11 years older than Seaborg, the two men had many things in common. Lawrence was born and raised in a small town in South Dakota and attended the University of South Dakota, where a charismatic teacher steered him away from premedical studies to physics. After receiving a PhD at Yale in 1925, he was lured to Berkeley by a generous salary and a promise of plenty of time for research. He conceived of the cyclotron one evening in 1929 while leafing through an obscure journal in the Berkeley library. Gazing at a half-understood drawing, he suddenly thought how he could use circular magnets and a rapidly fluctuating electrical field to accelerate charged particles to speeds never achieved before. He immediately knew the significance of what he'd found (he received the Nobel Prize for the discovery 10 years later). The morning after his evening in the library, encountering the wife of another faculty member on a campus walkway, he told her, "I'm going to be famous."

On the theoretical side, the Berkeley physics department revolved around a radically different man. J. Robert Oppenheimer was the son of wealthy, nonobservant Jewish parents who lived on the Upper West Side of Manhattan. A polymath in the elite private school he attended, he read the classics in Greek and Latin, studied French poetry, and was tutored by the curator of the American Museum of Natural History—all in high school. He studied chemistry, physics, philosophy, and mathematics at Harvard and then physics in Europe,

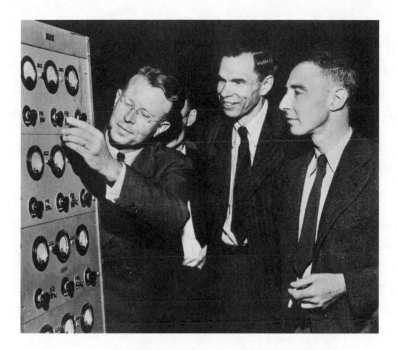

Ernest Lawrence, Glenn Seaborg, and Robert Oppenheimer
at the controls of Lawrence's University of California, Berkeley,
cyclotron. The men were at the center of America's effort to
build atomic bombs. *Courtesy of the US Department of Energy.*

receiving his PhD after an oral examination administered by the German physicist James Franck. As Franck said afterwards, "I got out of there just in time. He was beginning to ask me questions." During high school and afterwards, Oppenheimer had spent several summers on a ranch in New Mexico, and his love for the West was a factor in his arranging for joint positions at the California Institute of Technology and at Berkeley, starting in 1929, when he was just 25 years old. In both institutions, he was surrounded by a contingent of graduate students who tended to mimic his mannerisms, including the strange humming he made between paragraphs of impeccably crafted prose. They were sometimes called the "nim nim boys."

In the midst of all this scientific talent, Seaborg was at times overwhelmed. "I couldn't get over the feeling that I'd been plucked from the minor leagues and put on a major league all-star team. The world is filled with talented prospects who can't hit the curveball—would I turn out to be one of them?" He solved the problem with midwestern good sense: he resolved to work as hard as he possibly could. "That has proved to be the secret of whatever success I've had, if you call such a pedestrian notion as hard work a secret. Looking back, I can say that my whole life I've been surrounded by people who are brighter than I am, and I've done my best to take advantage of having them to work with."

Seaborg threw himself into graduate school life. He took courses, attended seminars led by Lewis, Lawrence, and Oppenheimer, taught an introductory chemistry lab to make money, and sought out an advisor whom he thought would give him lots of freedom to pursue his own path. His education at UCLA had given him a solid grounding in both chemistry and physics, and the intersection of physics and chemistry is where the excitement was at Berkeley in the 1930s. He got to know the physicists in the building at Berkeley that housed the cyclotrons—known as the Radiation Laboratory or Rad Lab. It was so full of radio waves, Seaborg recalled, you could light an electric bulb by touching it to any metal surface. He was an energetic and enthusiastic young graduate student eager to make his mark on the world.

●

IN 1934, THE SAME YEAR that Seaborg started at Berkeley, more remarkable scientific news emerged from Europe. Since the end of the 19th century, scientists had known that the heaviest elements found in nature, like radium and uranium, are unstable. Their nuclei emit subatomic particles like electrons and alpha particles (consisting of two protons and two neutrons) along with high-energy packets of light known as gamma rays, which scientists had collectively termed radiation. In the process, heavy elements gradually change from one element to another until they find a stable configuration of protons and neutrons.

This, too, makes sense. The more protons an atom has, the more fiercely they repel each other. The heaviest elements have lots of neutrons to hold their protons together—the most common isotope of uranium has 146 neutrons compared with 92 protons. But even a superabundance of neutrons is not enough to render heavy elements completely stable, and eventually they decay to lighter elements.

In January 1934 the husband and wife team of Irène and Frédéric Joliot-Curie (they hyphenated their last names when they were married) made an astonishing discovery. *All* elements, it turns out, have isotopes that are radioactively unstable. Bombarding any element with subatomic particles can create nuclei with unstable combinations of neutrons and protons that do not exist in nature. Physicists had assumed that such combinations would immediately fall apart. Instead, these oddball collections of protons and neutrons stick together, but just for a while. Over time, just like heavy elements, these unstable isotopes of light elements decay to new combinations of protons and neutrons until they find a stable configuration.

The discovery of what the Joliot-Curies termed artificial radioactivity was just as momentous as the discovery of the neutron. Suddenly, the universe consisted of much more than stable light elements and radioactive heavy elements. By bombarding stable elements with subatomic particles, physicists could create radioactive isotopes, or radioisotopes, with new, unusual, and highly useful properties. It was as if the surface of reality had gone transparent and revealed a shadow world beneath.

○

THE DISCOVERY OF ARTIFICIAL RADIOACTIVITY made Lawrence's cyclotron at Berkeley into a scientific goldmine. By bombarding different elements with fast-moving particles accelerated by the cyclotron, the Berkeley physicists could produce a virtually unlimited number of new and previously unknown radioisotopes.

With his background in both chemistry and physics, Seaborg was perfectly positioned to join the radioisotope hunters at Berkeley. Still, as a wet-behind-the-ears graduate student, he needed a break to get into the field. One day in April 1936, he was walking between the physics and chemistry buildings when a physicist named John Livingood stopped him. Livingood had been bombarding a target of tin in the cyclotron, seeking to create radioactive elements slightly heavier and slightly lighter than tin. Would Seaborg be interested in chemically separating the newly created elements from the irradiated target?

This was just what Seaborg needed. It would put him in direct contact with the work going on in the Rad Lab, an opportunity usually denied young graduate students. And once he started working on one project, he knew he would be caught up in others. He quickly set up a small lab in the physics building and bootlegged the chemicals and equipment he would need from the chemistry department. When Livingood brought him the irradiated tin, he dissolved the target in acid and added chemicals that combined first with the tin and then with indium (the element just before tin) and antimony (the element just after). "Deuteron-Induced Radioactivity in Tin," which was published a few months later in *Physical Review*, was Seaborg's first published scientific paper.

Over the next few years, Seaborg, Livingood, and their colleagues at Berkeley discovered dozens of radioactive isotopes, including several of the most important radioisotopes used in medicine and industry today. Cobalt-60 is used to treat cancer and sterilize medical instruments. Technetium-99 is used to image the liver, lungs, brain, and other organs. Iodine-131 is used to diagnose and treat thyroid problems. Seaborg later claimed that this last isotope was the most significant of all the ones he helped discover. Twenty-five years after its discovery, iodine-131 prolonged his mother's life by a decade when she contracted hypothyroidism, a disease similar to the one that had killed her sister.

Seaborg earned his PhD in 1937 for a thesis on the interactions of neutrons with lead. He then went to work for Lewis as a research assistant while continuing his own research in the mornings, during lunch breaks, and in the late afternoons and evenings. Seaborg was becoming more and more interested in the odd behavior of uranium and other heavy elements when they were bombarded by subatomic particles. They were giving off strange signals that were hard to interpret. Something interesting was going on in their nuclei, but no one could figure out what. Seaborg began spending more time with the physicists in the Rad Lab.

It was a backbreaking schedule, but Seaborg also took time to have fun. One night he and fellow chemist Willard Libby, who would win a Nobel Prize in 1960 for his development of carbon dating, went out with two other Berkeley colleagues and a young chemist named Henry Taube to celebrate Taube's new PhD. After going to Trader Vic's on San Pablo Avenue, where they drank Zombies to celebrate Taube's achievement, they went to a Chinese nightclub on Tenth Street in Oakland. When they entered the club, the new doctorate holder, who would eventually win a Nobel Prize for his work on how electrons behave in metals, pitched headlong onto the floor, "the result of too many Zombies," Seaborg recalled. "I wonder if any of the people in that night club who witnessed our arrival would have believed that three of us were to win Nobel Prizes."

Chapter 2

THE CHAIN REACTION

ON TUESDAY, JANUARY 31, 1939, LUIS ALVAREZ, A 27-YEAR-OLD physicist working at Berkeley's Rad Lab, was reading the *San Francisco Chronicle* while getting his hair cut at the student union building. On the second page he came across the following article:

200 Million Volts of Energy
Created by Atom Explosions
WASHINGTON, Jan. 29 (AP)—American scientists heard today of a new phenomenon in physics—explosions of atoms with a discharge of 200,000,000 volts of energy.

The article, though somewhat confused on the details, explained that German chemists had been bombarding uranium atoms with neutrons and had discovered that the atoms were splitting roughly in half to form much smaller atoms, a process they later termed fission. When the atoms split, they gave off immense amounts of energy— far more than the energy released in chemical reactions. Scientists quoted in the article, including the famous Italian physicist Enrico Fermi, said that the finding was comparable in significance to the discovery of radioactivity.

Alvarez jumped from the barber chair, his hair half cut, and ran up the hill to the Rad Lab. There he encountered fellow Rad Lab physicist Philip Abelson, who also had been trying to understand the odd behavior of uranium when it was bombarded with neutrons in the cyclotron. "I have something terribly important to tell you," Alvarez said. "I think you should lie down on the table." Abelson lay down next to the cyclotron's control panel. Alvarez told him the reason he and other scientists around the world had been getting such strange results from the bombardment of uranium: the uranium atoms were splitting into pieces.

"When Alvarez told me the news, I almost went numb," Abelson recalled. Like everyone else, he had assumed that the signals he was measuring from the bombarded uranium were coming from elements slightly heavier or slightly lighter than uranium. But the signals had been coming from much smaller atoms all along. Abelson realized that he "had come close but had missed a great discovery."

Seaborg heard the news that evening at the weekly meeting of young physicists and chemists to discuss recently published papers. Some of the attendees didn't believe it. That's not what they had learned about atoms. Bombarding elements in the cyclotron produced elements just before and just after that element on the periodic table—it didn't split the atoms into much smaller atoms. As Seaborg later put it, "If you hit a car-size boulder with a pick, you may chip off a piece, but you won't split the boulder into two halves." But now the German chemists were saying that hitting uranium atoms with neutrons was doing just that. It was splitting the atoms into barium, with 56 protons, and krypton, with 36.

Despite the others' skepticism, Seaborg knew immediately that the Germans were right. For years, physicists and chemists had been puzzled by the variety of signals they were getting from irradiated uranium. The splitting of uranium atoms into smaller atoms, each of which produced its own radioactive signals, explained the results perfectly.

After the journal club, Seaborg was distraught. "I walked the streets of Berkeley for hours with the news whirling around my head," he later remembered. "My mood alternated between exhilaration at the exciting discovery and consternation that I'd been studying this field for years and had completely overlooked the possibility of this phenomenon—and missed a chance for an astounding discovery."

Other scientists were equally dismayed. When the physicists at the Rad Lab had bombarded uranium in the cyclotron, they had noticed jolts of energy on their radiation detectors. But at the time they had decided that their detectors were malfunctioning and had turned them off rather than searching for the source of the signal. The Joliot-Curies had missed fission, too, refusing to consider the possibility that uranium was splitting into smaller atoms. Even Enrico Fermi's team at the University of Rome, which in 1934 had pioneered the bombardment of uranium with neutrons, had overlooked the fissions occurring in their experimental apparatus. After the German announcement, physicists everywhere realized that the signals had come from fissioning uranium atoms.

After a few days of self-recrimination, Seaborg came to terms with the discovery. "What an exciting specialty I'd chosen," he decided. "What great fortune to be in a field with so much work to be done."

JUST A FEW WEEKS EARLIER, in the lobby of the King's Crown Hotel in New York City, two émigré European scientists living in the United States met each other for the first time. If Seaborg's decision to go into science was one of the keys to the launch of Hanford, that meeting was the other.

One of the two men was Enrico Fermi—at that point, with the possible exception of Albert Einstein, the best-known physicist in the world. Fermi was perhaps the last great physicist who was equally skilled as a theoretician and an experimentalist. Earlier in the decade, as a young professor at the University of Rome, he had developed a theory that explains radioactive decay, proposing that it results from a force in nature different from gravity or electromagnetism. He and his colleagues at the university were also the first to use neutrons to convert elements into radioactive isotopes, discovering in the process that slowly moving neutrons are especially effective in reconfiguring the neutrons and protons in nuclei. Fermi was a star in Italy, surrounded by skilled colleagues, lauded by the media, honored by Mussolini.

But Fermi's wife Laura, the smart, beautiful, and elegant daughter of an admiral in the Italian Navy, was Jewish, which made their two children Jewish. By 1938, Mussolini's growing anti-Semitism had made it impossible for the Fermis to remain in Italy. That year, Fermi won the Nobel Prize in Physics for his work on slow neutrons. He and his family used the trip to Stockholm, plus an eagerly offered visiting professorship at Columbia University, as a way to depart from Italy without raising suspicions that they were leaving for good. After the award ceremony they traveled to England and then to New York City, where they took rooms in the King's Crown while looking for an apartment to rent. They were right to flee. During the war, Laura's father was sent to a concentration camp and never heard from again.

The other man in the lobby of the King's Crown was Leo Szilard, the oldest child of an assimilated Jewish upper-middle-class family in Budapest. During graduate school in Berlin, Szilard had written a dissertation on the connection between information theory and thermodynamics that attracted widespread attention, including from Albert Einstein. But he was never interested in becoming an academic scientist. Szilard's interests were wide ranging, from physics and biology to history, literature, and politics, and he rarely stuck with a single subject for long. Instead, he drifted from topic to topic, from city to city, and eventually from country to country. He stitched together projects and sources of income, often as a part-time or short-term researcher at universities where he knew someone with influence. He lived in hotels, in faculty residences, and, when worse came to worst, with friends. He never seemed to care much about his circumstances. For Szilard, only ideas mattered.

In 1933, Szilard was living at the Imperial Hotel in the Bloomsbury neighborhood of London and running an organization that he had cofounded to help refugee scholars escape from Nazi Germany. He was always adept at forecasting future developments and predicted, well before most people, that the rise of Hitler would lead Germany to war. On September 12, he read an article in *The Times* in which the physicist Ernest Rutherford was quoted as saying that anyone predicting the generation of energy from atomic nuclei was "talking moonshine." The comment irritated Szilard. "How can anyone know what someone else might invent?" A few days later he was going for a walk in the neighborhood when he stopped for a red light at a crosswalk near the hotel. When the light turned green and he stepped off the curb, he thought of something. If an element could be found that released two neutrons when it was hit by one neutron, that element could create a chain reaction and release nuclear energy, thereby proving Rutherford wrong. Fission would not be discovered for another five years, and Szilard did not think right away of uranium as a chain reaction candidate. Still, in stepping off that curb, he conceived of one of the most important scientific ideas of the 20th century.

In science, new ideas often occur to more than one person at about the same time. They are "in the air" because previous developments point in their direction. That was not the case for the chain reaction. For the next five years, of all the scientists in the world, only Szilard thought deeply about the possibility. When he mentioned the idea to others, they mostly dismissed it as unlikely or impractical. At the same time, Szilard did not mention it to many people, because he immediately recognized the idea's danger. If a chain reaction were possible, and if German scientists learned how to create one, they might be able to build a bomb so powerful that Germany could conquer the world.

Early in 1938, tiring of Europe and its politics, Szilard moved to New York City, where he took a room at the King's Crown. He spent much of that year traveling from place to place, meeting with people he knew in various academic and industrial laboratories and trying to get them interested in the chain reaction. But most were uninterested or unable to help. By the end of 1938, he was back at the King's Crown and ready to give up.

Szilard met Fermi just as the news of fission was making its way to America. Suddenly, the chain reaction that Szilard had pursued for years seemed within reach. If uranium gave off neutrons when it fissioned, those neutrons could cause more fissions, which in turn would give off more neutrons. But did fissions produce neutrons? Within a few weeks of their meeting, both Fermi and Szilard were working on the problem. Fermi had struck up a partnership with a young graduate student at Columbia named Herbert Anderson, just 24 years old, a scholarship student from the Bronx who had been studying electrical engineering before getting interested in physics. Szilard, meanwhile, had convinced a Canadian-born physicist at Columbia named Walter Zinn, 32 that year, the son of a factory worker, to collaborate with him on fission research. Before long, both groups had found the extra neutrons, as had several other groups in the United States and Europe. For Szilard, the discovery was vindication that the idea he'd had while crossing the street in London was correct. Yet he also knew that the discovery brought the world one step closer to an atomic bomb. The evening he and Zinn discovered the extra neutrons produced by fission, he later wrote, "there was very little doubt in my mind that the world was headed for grief."

The discovery that uranium releases extra neutrons raised another question: why don't chain reactions occur in uranium ore? This question, too, was answered in the first few months of 1939. Uranium ore consists almost entirely of two isotopes—one with 92 protons and 146 neutrons, known as uranium-238, and one with 92 protons and 143 neutrons, or uranium-235. Theory suggested, and experiments confirmed, that only uranium-235 was fissioning when it was hit with neutrons. But uranium-235 makes up only about seven-tenths of one percent of uranium ore. That's too low to support a chain reaction.

Physicists also discovered that uranium-235 was splitting not just into barium and krypton but into many different pairs of atoms—

xenon and strontium, and iodine and yttrium, and cesium and rubidium—into more than 20 different elements altogether. These fission products, as they're called, tend to be highly radioactive. That's because they have too many neutrons for the number of protons they contain. Uranium atoms, to keep their 92 protons together, have proportionately more neutrons in their nuclei than do lighter elements. As a result, when uranium atoms split, their fission products have a sort of neutron sickness—they have too many neutrons to be stable. To get back into balance, these fission products begin to convert neutrons in their nuclei into protons while releasing electrons and gamma rays. In this way, the fission products start crawling their way down the list of elements, away from hydrogen and toward uranium, until their neutron sickness is resolved.

When physicists realized that only uranium-235 fissions in the presence of neutrons, they immediately saw one way to create a chain reaction. If enough uranium-235 could be extracted from natural uranium ore, the neutrons from fissioning atoms would cause other atoms to fission in an ever-growing cascade. If this reaction could be controlled, uranium-235 could provide virtually unlimited quantities of energy, whether for electricity generation, propulsion, or other purposes. If the reaction were uncontrolled, and if it occurred quickly enough, the result would be a bomb of unprecedented power.

But isolating uranium-235 from uranium ore was going to be extremely difficult. Because the two uranium isotopes are different versions of the same element, they behave exactly the same chemically. The only way to separate them is on the basis of their weights. But they differ in weight by only about one percent, which provides very little traction for a separation process. At the time, several technologies existed that might do the job, but they would have to become much more efficient and then be applied on a huge scale. As the Danish physicist Niels Bohr famously remarked, "It would take the entire efforts of a country to make a bomb."

But Szilard was not going to give up on the chain reaction. Maybe a way could be found to arrange natural uranium ore so that a chain reaction could be produced some other way. And as more was learned about fission, he began to glimpse an intriguing possibility.

The neutrons emitted by a fissioning uranium-235 nucleus travel very quickly—at about one-fifteenth the speed of light. When a neutron traveling at this speed hits a uranium-238 nucleus, one of two things usually happens. The neutron can bounce off the nucleus and head in another direction. Or it can be absorbed by the nucleus, creating uranium-239. In fact, this capture of neutrons by uranium-238 is the real reason why natural uranium ore does not explode. The absorption of neutrons by the heavier isotope of uranium squelches any possible chain reaction.

But if the fast-moving neutrons from fission are slowed down, something else happens. Slow neutrons are less likely to be captured by uranium-238 atoms. Instead, they bounce harmlessly off uranium-238 until they find a nucleus of uranium-235 to fission. That's the trick to achieving a chain reaction with natural uranium, Szilard realized. If some sort of moderator could be found that would slow down the fast neutrons from uranium-235, fewer neutrons would be captured by uranium-238, leaving more than enough to produce a chain reaction.

But how could neutrons be slowed down? Remarkably, Fermi had answered this question just a few years before. High-speed neutrons slow down quickly when they bounce off the nuclei of light atoms. So to find a good moderator, Szilard and Fermi, working together now as each recognized the potential of chain reactions, began making their way down the list of elements. Hydrogen atoms, with their single proton, work best at slowing down neutrons. But they occasionally absorb neutrons, slowing the chain reaction. The next heavier element, helium, does not absorb neutrons, but suspending uranium ore in a container of helium gas did not seem immediately practical. Lithium, with three protons, is a strong neutron absorber, so that would not work. Beryllium, with four protons, does not absorb many neutrons, but it is highly toxic when inhaled. Boron, with five neutrons, absorbs neutrons like crazy. But what about carbon, with six protons and six neutrons? It captures neutrons at one-hundredth the rate of hydrogen. And a source of concentrated carbon was readily available: graphite, the "lead" in lead pencils.

By the summer of 1939, Szilard had convinced himself that graphite would work. "It seems to me now that there is a good chance that carbon might be an excellent element to use" as a moderator, he wrote in a letter to Fermi. "I personally would be in favor of trying a large-scale experiment with a carbon-uranium-oxide mixture if we can get hold of the material." But getting hold of the material wasn't going to be easy. A few days later, Szilard visited the National Carbon Company on East Forty-Second Street to ask about the cost of acquiring graphite blocks. It could be done, but the graphite had to be extremely pure, and that would be expensive. Fermi and Szilard also needed a large amount of graphite. A slow neutron bouncing around in a large block of graphite containing embedded masses of uranium—which is the design that Szilard and Fermi gradually worked out over the next couple of years—can do one of two things. It can find an atom of uranium-235 to split. Or it can reach the edge of the graphite and escape, in which case it can no longer fission uranium. For a chain reaction to occur, a mass of graphite and uranium would have to be large enough for a neutron to find a uranium-235 atom before it was lost to the surrounding space.

In the spring of 1940, Columbia University received $6,000 from the federal government to cover the purchase of a large quantity of uranium ore and extremely pure graphite. Fermi, Szilard, Anderson, Zinn, and a steadily growing research team immediately began to experiment with the materials Szilard bought with the new funds. Using bench saws and planers, they machined the graphite into four-inch by four-inch by twelve-inch blocks. They then stacked the blocks in layers until they had a column of dusty black graphite. The physicists at Columbia "started looking like coal miners," Fermi recalled, "and the wives to whom these physicists came back tired at night were wondering what was happening." The experimenters put neutron sources under the graphite blocks to measure how quickly the neutrons slowed down and whether they were being absorbed by remaining impurities in the graphite. They then put uranium in the midst of the graphite blocks. Sure enough, when neutrons from the bottom of the column encountered the uranium, fissioning uranium-235 atoms generated new neutrons. Fermi and Szilard were not yet calling this device by the name its much larger descendants would acquire—a *reactor*. They called it what it was—a *pile*.

Chapter 3

ELEMENT 94

EVEN AS FERMI WAS BUILDING HIS PILES AT COLUMBIA UNIVERSITY, Seaborg was beginning the experiments that would link the work of the two men forever. Since 1939, a physicist at Berkeley named Ed McMillan had been studying the atoms produced when uranium fissions, and one of them had caught his attention. Because of the energy generated when uranium splits, most fission products fly away from the site of the fission at high speeds. This one didn't. It was radioactive, just like a fission product. But it stayed close to the place where it was created. McMillan began to suspect that it wasn't a fission product at all.

For help he called on Phil Abelson, the Rad Lab employee who had been so distraught at barely missing the discovery of fission. Abelson had a background in chemistry, and he soon was able to show that the atom McMillan was studying had chemical properties different from those of any other known element. Only one conclusion could be drawn. It must be a new element. The most plausible explanation was that the uranium-238 was capturing a neutron from fissioning uranium-235 atoms. That would make it uranium-239, an isotope of uranium with 92 protons and 147 neutrons. But this isotope of uranium was evidently unstable. It appeared to be converting one of its neutrons into a proton, just as fission products do. The result would be an element with 146 neutrons and 93 protons, a *transuranic* element, an element never before seen on Earth.

As the discoverers of the new element, McMillan and Abelson

had the privilege of naming it. Uranium had been named after the planet Uranus. McMillan and Abelson therefore named their new element neptunium, after the next planet out from the sun.

Seaborg had been following McMillan's work closely—he would even pester McMillan with questions when the two met in the shower room of the faculty club. Then, in the fall of 1940, McMillan suddenly moved to MIT. Scientists in the United States were gearing up for an event that many people thought was inevitable—America's entry into World War II. Instead of studying radioactive elements, McMillan spent the next few years developing radar.

When Seaborg learned that McMillan was giving up his work at Berkeley, he wrote a letter asking if he and his colleagues could continue McMillan's research. McMillan wrote back to say that he would be happy to turn the work over to Seaborg's group.

Seaborg and his colleagues at Berkeley had reason to believe that the neptunium found by McMillan and Abelson was also undergoing radioactive decay. If so, it would convert another one of its neutrons into a proton. The result would be an element with 94 protons—another transuranic element unknown in nature. Because of the way its nucleus was configured, this element would probably be much more stable than either uranium-239 or neptunium. And if this was true, element 94 could be a very interesting element indeed.

WELL PAST MIDNIGHT on February 24, 1941, a storm was roiling the waters of San Francisco Bay as Seaborg and Art Wahl, a 23-year-old graduate student, prepared for the decisive experiment in Room 307 of Berkeley's Gilman Hall. Many years later, when Room 307 of Gilman Hall was being designated a National Historic Landmark, Seaborg said that "a less significant or historical looking room hardly existed on the campus of the University of California." It had a small sink set in a heavily stained stone countertop. Discolored iron pipes nearly filled the space between the countertop and a glass-fronted cabinet filled with apparatus and chemicals. The room was on the top floor of Gilman Hall, and a set of glass doors led to a small balcony notched into the red clay roof tiles of the building. The chemical hood in the room did not have a fan, and when Seaborg and Wahl wanted to do a particularly noxious experiment they moved their equipment to the balcony. But on this particular night rain was spattering the west-facing doors, and Seaborg and Wahl braved the fumes of their experiment inside.*

About two months earlier, Wahl, who had come to Berkeley about a year earlier after majoring in chemistry at Iowa State University, had spread a uranium paste onto a copper plate. He and Seaborg then placed the plate in the target area of the 60-inch cyclotron at Berkeley. After the plate had undergone four hours of bombardment, Wahl, back in Gilman Hall, scraped the powder off the plate with an ice pick. He dissolved the scrapings in hot nitric acid and was alarmed to see the solution turn green. The uranium was doing something it should not be doing. But then he remembered—the copper plate. Some of the copper must have come off with the irradiated uranium.

He took a bottle of cerium fluoride from the shelf and mixed some into the solution. The white powder caused the uranium, the neptunium, and any elements derived from neptunium to form solid particles in the liquid. Wahl separated this precipitate from the solution with a filter and attached it to a piece of cardboard.

For the next several weeks, Seaborg and Wahl plied their sample of bombarded uranium with chemicals, trying to figure out how to isolate the new element they suspected they had made. When they wanted to know what elements were present in a solution or a precipitate, they took it two doors down the corridor to Room 303 in Gilman Hall, where their colleague Joseph Kennedy had built an ingenious set of radiation detectors. Every radioactive isotope gives off a dis-

---

* As is usually the case in science, Seaborg worked with several collaborators. Labeling him the "discoverer" of element 94 requires a judgment call about the extent to which he was responsible for the direction of the research.

tinct set of signals—for example, electrons and gamma rays with a particular range of energies. By measuring these signals in his detectors, Kennedy could tell what radioisotope each sample contained.

But Seaborg and Wahl had a problem. To isolate their suspected new element from the irradiated uranium, they needed a chemical that would combine only with that element and not with any other, but they couldn't find a compound strong enough to do the job. Finally they asked Berkeley chemist Wendell Latimer. He recommended the strongest combining agent he knew, peroxydisulfate, a compound in which two sulfur atoms brandish eight oxygen atoms like spikes on a cudgel. Late in the evening on Sunday night, February 23, Wahl dissolved some of the irradiated uranium in acid, added some peroxydisulfate, and let the solution stand for about half an hour. He then added hydrofluoric acid to the solution, which caused the remaining uranium and neptunium to form solids that he separated from the solution with a filter. But Kennedy's detectors showed that the new element remained in the solution, meaning that it "can be separated from all the known elements," as Seaborg wrote in his journal.

Seaborg later recalled stepping onto the balcony of Gilman Hall after the momentous experiment. Ahead, the sky over San Francisco Bay reddened with the rising sun.

"I was a 28-year-old kid and didn't stop to ruminate about it," he later told an *Associated Press* reporter. "I didn't think, 'My God, we've changed the history of the world.'"

But Seaborg and Wahl had discovered an element that would change the history of the world.

●

EVEN BY THE STANDARDS of the Berkeley chemistry department, Seaborg made for a strange sight as he struggled up the steps of the chemistry building. In addition to his lab coat he wore heavy lead-impregnated gloves and goggles. At the end of a long pole he carried a lead bucket. He was trying to protect himself from the radioactivity being emitted by the fission products in the bucket. Whether he succeeded is questionable. Years later, Room 307 in Gilman Hall had to be thoroughly decontaminated to temper the radioactivity emanating from its walls, floor, and countertops.

Seaborg and his colleagues were conducting one of the most consequential experiments of the nuclear age. Could the new element that they had created be used to build atomic bombs? The idea seemed plausible. The reason uranium-235 fissions while uranium-238 does not is that the former has an odd number of protons and neutrons while the latter has an even number. Like dancers, protons and neutrons like to pair up in nuclei, which makes even-numbered nuclei less likely to fission from exposure to neutrons than odd-numbered nuclei. But the radioisotope Seaborg's group had created, like uranium-235, had an odd number of nuclei—94 protons and 145 neutrons, for a total of 239. Furthermore, that isotope of element 94 is relatively stable and remains largely unchanged for thousands of years. If their new element fissioned like uranium-235 fissions, it, too, could serve as the fuel for atomic bombs.

In the last week of February 1941, Seaborg and his colleagues had placed about two-and-a-half pounds of uranium in the target of the cyclotron and had bathed the uranium in neutrons. By the time they were done, the sample—200 times larger than the sample used to discover element 94—was filled with fission products and dangerously radioactive.

On Monday, March 3, 1941, Seaborg and his colleague Emilio Segrè transported the irradiated uranium from the cyclotron to Gilman Hall, where the physics department had made them an extraction apparatus that they could operate by remote control. They dissolved the irradiated uranium in two liters of ether and separated out the remaining uranium, leaving behind neptunium, their new element (which was continuously being created by the decay of neptunium), and lots of fission products. They placed this solution in a tube that they carried to a nearby laboratory with a large centrifuge. There they spun the tube to separate element 94 from the lighter fission products. Six times they repeated this procedure,

When Room 307
of Gilman Hall was
dedicated as a
national landmark
in 1966, Art Wahl
and Glenn Seaborg
posed with the
sample of plutonium
that they produced
there in March 1941.
*Courtesy of the US
National Archives
and Records
Administration.*

each time getting a purer sample of their new element. Finally they poured the purified sample into a platinum dish about the size of a dime. They labeled it Sample A and stored the sample in a cigar box.

The crucial test came on March 28. Seaborg and his colleagues placed their sample of element 94 in the path of neutrons being generated by the cyclotron. Almost immediately a detector began picking up the unmistakable signals of fission. More testing revealed that the new element generated even more neutrons when it fissioned than did uranium-235. That meant it could produce even more powerful atomic bombs.

It took a while for the implications of this discovery to sink in, but they were profound. Seaborg and his colleagues had discovered a way to convert the common isotope of uranium, which could not support a chain reaction, into a previously unknown element that could. In a stroke, they had multiplied by a hundredfold the amount of energy that could be extracted from uranium. Furthermore, they had shown that they could use chemistry to separate this new element from other elements. They did not need to rely on the small difference in weight between uranium-235 and uranium-238 to produce bomb-making material. They had found a new and much easier way to make atomic bombs.

○

BY THE SPRING OF 1941, Seaborg and his colleagues were no longer talking about their discoveries in public—or even with most other scientists. As soon as scientists in the United States realized that certain radioactive substances might be used to build atomic bombs, they quit publishing their results in the scientific literature, afraid that anything they said could help German scientists develop a bomb. But physicists in Germany immediately saw what was happening and drew the obvious conclusion: American physicists must be working on nuclear weapons. Physicists on both sides of the conflict concluded that they were in a race to see which country could build atomic bombs first.

To maintain secrecy, Seaborg and his colleagues initially referred to their new element simply as 94, for the number of protons it contained. But they knew it would eventually need a real name. At first, they mistakenly thought that it would be the heaviest element ever discovered and considered such names as *extremium* and *ultimium*. "Fortunately, we were spared the inevitable embarrassment that one courts when proclaiming a discovery to be the ultimate in any field," Seaborg later reflected.

They decided to follow precedent. The planet Pluto had been discovered in 1930, when Seaborg was an undergraduate at UCLA. After Uranus and Neptune had been honored, Pluto seemed the logical next step. Seaborg's group briefly considered the name *plutium*, but they thought that *plutonium* sounded better. Much was made in later years about Pluto being the god of the underworld, but Seaborg later professed to be unaware of the connotation. "I was unfamiliar with the god or why the planet was named for him," he said. "We were simply following the planetary precedent."

Chapter 4

THE DECISION

DESPITE THE SECRECY SURROUNDING PLUTONIUM, WORD OF SEABORG'S achievement spread quickly through the small network of scientists and policymakers who were thinking about atomic bombs. Almost immediately, they saw something that seemed too fortuitous to be true.

The Berkeley scientists had discovered a way to create a bomb-making material—plutonium-239—by bombarding uranium-238 with neutrons. But neutrons were hard to generate in quantity: cyclotrons could never make enough to produce atomic bombs.

However, Fermi and Szilard, at Columbia University, were working on a device—a nuclear reactor—that could generate plenty of neutrons beyond those needed to keep a chain reaction going. Furthermore, the isotope that needed to be bombarded with neutrons, uranium-238, was right there in the reactor's uranium ore! A nuclear reactor would make plutonium simply by operating. It was like driving a car that created more gasoline the farther it went.

Meanwhile, Seaborg had shown that plutonium could be separated from irradiated uranium using conventional chemical processes. If all these processes were combined, a fuel for atomic bombs could be made just with nuclear reactors and chemical processing plants.

●

GOVERNMENT OFFICIALS HAD KNOWN almost since the discovery of fission that atomic bombs might be possible if enough uranium-235 could be separated from natural uranium ore. But they did not do much to follow up on the possibility until Seaborg and his colleagues discovered plutonium in the spring of 1941, and one man was at the forefront of that change. Vannevar Bush was, in the estimation of biographer G. Pascal Zachary, the "engineer of the American century." Educated at Tufts College and the Massachusetts Institute of Technology, Bush spent the first 20 years of his professional life doing research, teaching, and rising through the ranks of the MIT administration. Known as Van to his friends—he considered his first name "a nuisance"—he developed techniques for detecting submarines during World War I, built a forerunner of today's electronic computers, and cofounded the company now known as Raytheon. But he wanted to do more than invent. He wanted inventions to be used.

In 1939, Bush moved from Massachusetts to Washington, DC, to become president of the Carnegie Institution of Washington. From his new office just eight blocks north of the White House, Bush was ready to do what he thought he could do best: shepherding inventions from America's labs into the federal government, and especially into the military. But he also knew that "you couldn't get anything done in that damn town unless you organized under the wing of the President." Through a Carnegie trustee who was Franklin Roosevelt's uncle, he arranged a meeting with the president on June 12, 1940. There he gave Roosevelt a memo proposing a National Defense Research Committee that would "correlate and support scientific research on mechanisms and devices for warfare." After less than 15 minutes of discussion, Roosevelt approved the plan. Bush would remain independent of government and continue to be paid through the Carnegie Institution, but he now had a direct line to the president and funds from the White House budget to prepare new technologies for the military.

Many technologies were on the verge of making wartime contributions—radar, new kinds of submarine detectors, electronic devices to set off explosives as they neared their targets. But one loomed above the rest—what Bush called "this uranium headache." As he told a friend in the spring of 1941, "I wish that the physicist who fished uranium in the first place had waited a few years before he sprung this particular thing on an unstable world." Rumors were emerging from Europe that German physicists were working on fission, and Bush acknowledged that he was "scared to death" of a Nazi bomb. But isolating enough uranium-235 to make a bomb looked like it would take years, if it could be done at all, and the news about plutonium was just starting to emerge from Berkeley.

As would happen often in the next few years, Bush turned for advice to a committee. The National Academy of Sciences, chartered by President Lincoln in 1863 during the darkest days of the Civil War, was mostly an honorific society for esteemed scientists. But its Act of Incorporation also states that "the Academy shall, whenever called upon by any department of the Government, investigate, examine, experiment, and report upon any subject of science or art." In the spring of 1941, Bush gave the Academy the most important assignment it had ever received: to review the possible military uses of fission.

The Academy created a six-member committee, chaired by Arthur Compton, dean of science at the University of Chicago, to conduct its review. In 1927, Compton had won a Nobel Prize for his discovery of what is known as the Compton effect, which demonstrated that light can behave like a particle as well as a wave. He was also an intensely religious man: "God can have no quarrel with a religion which postulates a God to whom men are as His children," he once told a reporter from *Time* magazine, which put him on the cover of its January 13, 1936, edition. Yet he had no qualms about leading a committee charged with investigating the engines of war: "As long as I am convinced, as I am, that there are values worth more to me than my own life, I cannot in sincerity argue that it is wrong to run the risk of death or to inflict death if necessary in the defense of those values."

The Academy committee, which included Berkeley's Ernest Lawrence, submitted its report on May 17. The committee noted that sep-

Arthur Compton (left) was chair of the National Academy of Sciences committee that advised Vannevar Bush (right) about the prospects for developing atomic bombs. *Corbis via Getty Images.*

arating large quantities of uranium-235 from uranium ore would probably take at least three to five years, even under optimistic assumptions. But there was another possibility, the committee pointed out. If a nuclear reactor could be built, and if plutonium could be chemically extracted from the irradiated uranium such a reactor would produce, bombs using plutonium could be built "within twelve months from the time of the first fission chain reaction," according to the committee. It recommended "strongly intensified" research on both isotope separation and plutonium production. "Within a half dozen years the consequences of such investigations may be crucial in determining the nation's military position."

After a few days of reflection, Bush found himself unsatisfied with the report. The pressure to do something more than study the problem was growing. Bush had been receiving secret information from a committee in England that also had been examining fission. It had concluded that an atomic bomb could be built within three years and that, as it stated in its own report, "the first side to perfect this scheme will gain a decisive and crushing victory." But Britain, holding out against the Nazis, with its factories straining to produce armaments, could not hope to undertake a massive bomb project. American scientists should take the lead, their British counterparts urged.

Bush was also beginning to realize how important atomic bombs would be after the war, regardless of whether they were used during it. The British nuclear program was well ahead of the American program. If the United States did not accelerate its work on fission, Britain could emerge from the war with the lead in nuclear technologies. Furthermore, the first country to build atomic bombs would gain tremendous political power—whether that country was the United States, Britain, Germany, or the Soviet Union. "I still shudder when I think what sort of a world it would have been if we had quit, and Russia had completed the job," Bush later wrote.

He asked the Academy committee to look at the issue again, but this time with additional members who could provide an engineering perspective on the project. The committee delivered its second report to Bush on July 11. That report was as skeptical as the first about the prospects for separating uranium-235, stating that the technologies being studied showed "little, if any, more promise of immediate or even early application than at the time of the previous report." But the committee, and especially Lawrence, was increasingly optimistic about the prospects for plutonium. "If large amounts of element 94 were available it is likely that a chain reaction with fast neutrons could be produced," Lawrence wrote in an appendix to the report. "In such a reaction the energy would be released at an explosive rate which might be described as a 'super bomb.' "

If isotope separation were the only route to a bomb, the Manhattan Project might never have happened. But Seaborg's discovery of plutonium in early 1941 provided a second option, and by that summer this second option was a powerful inducement. Even if isotope separation failed, plutonium production could work. And the availability of two routes to a bomb meant that Germany, which certainly had the scientific talent to discover plutonium, could get there two ways as well.

The next time Bush met with Roosevelt to discuss fission research was on October 9, 1941. By this time, Bush's mind was made up: "I knew that the effort would be expensive, that it might seriously interfere with other war work. But the overriding consideration was this: I had great respect for German science. If a bomb were possible, if it turned out to have enormous power, the result in the hands of Hitler might enable him to enslave the world. It was essential to get there first, if an all-out American effort could accomplish the difficult task." At the meeting with the president, Bush described the amount of material needed for a bomb, the cost of production plants, and how much time would be needed to build a weapon. He could not guarantee that the effort would be successful, but the science was promising and the implications of success profound. Again, Roosevelt quickly approved the plan—a greatly expanded research project was clearly warranted. But any steps toward production of bomb-making materials or of a bomb itself would require further approval from the three men at the meeting—Bush, Roosevelt, and vice president Henry Wallace—plus two others who would make pivotal decisions about atomic strategy throughout the war: Secretary of War Henry Stimson and Army Chief of Staff George Marshall.

Even with Roosevelt's okay, Bush wanted the blessing of his scientific advisors before initiating a massive research program. The same day he met with Roosevelt, he asked the National Academy of Sciences committee for a third assessment of fission. This report, delivered on November 6, 1941, was much less equivocal than the earlier two. "Within a few years the use of explosive fission may become the predominant factor in military action," the report stated. Regardless of whether atomic bombs used uranium-235 or plutonium, they would be weapons "of superlatively destructive power," the committee wrote. "This seems to be as sure as any untried prediction based upon theory and experiment can be."

○

NOW THAT BUSH HAD the president's support, he moved quickly to reorganize the government's fission program. He put Ernest Lawrence and Harold Urey, a physicist at Columbia University, in charge of overseeing the work on isotope separation—the former because he was convinced he could convert his cyclotrons into devices for doing the job. A planning board under the direction of Eger Murphree, a chemical engineer at the Standard Oil Development Company, started considering how to scale up laboratory-sized experiments to industrial-sized processes. And he put the University of Chicago's Arthur Compton, the chair of the National Academy of Sciences committee, in charge of research on the chain reaction and the production of plutonium.

At this point, the use of plutonium to construct an atomic bomb was still a long shot. Seaborg had discovered it less than a year earlier, and its basic properties and chemistry remained murky. Most people did not even know that a new element had been discovered, since Seaborg and his colleagues had kept their work secret. But Compton, after being proselytized by Lawrence, was a convert to plutonium. On Saturday, December 6, 1941, he met Bush and Bush's second-in-command at the National Defense Research Committee, Harvard University president James Conant, for lunch at the old Cosmos Club on Lafayette Square, just north of the White House. There the conversation turned to plutonium. Conant, an expert on acids and bases who had worked on the development of poison gases during World War I, remained highly skeptical about plutonium's prospects. Almost nothing was known about its chemistry, he pointed out. Who knew if it could be extracted from irradiated uranium? And doing so would be extremely difficult because of the intense radioactivity of the accompanying fission products.

Compton responded, "Seaborg tells me that within six months from the time the plutonium is formed he can have it available for use in the bomb."

Conant replied, "Glenn Seaborg is a very competent young chemist, but he isn't that good."

The next morning—Sunday, December 7, 1941—350 Japanese aircraft launched from six aircraft carriers attacked the US naval base at Pearl Harbor. More than 2,400 Americans were killed, and another 1,100 were wounded. All eight battleships in the harbor were damaged and four were sunk, along with several smaller ships. The next day the United States declared war on Japan; three days later Hitler declared war on the United States. America's isolation from the rest of the world had ended.

◦

THE NEXT MONTH, Seaborg was relieved of all his teaching responsibilities at Berkeley so he could work full-time on plutonium. The university had received $400,000 for bomb-related work, though much of that went to adapting Lawrence's cyclotron to separate uranium isotopes. Still, Seaborg now had the resources to begin addressing a long list of questions he had about plutonium: Can it fission without being hit by a neutron? How long does it take to decay? Are there better ways to separate it from other elements?

In March 1942, Compton's assistant came to Berkeley to tell Seaborg that he should plan on moving to Chicago. Compton had decided that all the work on plutonium should be concentrated in a single place, and he had chosen his own university as the location. Seaborg boarded the train for Chicago on April 17, 1942. He arrived two days later, at 9:30 in the morning. It was his thirtieth birthday.

Chapter 5

THE MET LAB

WHEN SEABORG STEPPED OFF THE TRAIN IN CHICAGO, THE TEMPERATURE
was 40 degrees—a shock to his California sensibilities. On the platform he read the headline of the Chicago *Sun*: "Tokyo Fears New Bombings; Reports Fires in Four Cities." It was a story about the Doolittle Raid of the previous day. After Pearl Harbor, President Roosevelt demanded that his military leaders retaliate against Japan as soon as possible. Organized and led by Jimmy Doolittle, a 43-year-old MIT PhD and stunt pilot who pioneered the concept of instrument flying, 16 B-25 bombers took off from the aircraft carrier USS *Hornet* and dropped high explosives and incendiary bombs on Tokyo, Yokohama, and several other cities, killing more than 50 people and injuring more than 400. The bombers could not return to the *Hornet* and land, so most of them crashed-landed in China after their crews bailed out. Remarkably, most of the 80 crew members survived and eventually returned to the United States. The raid provided a great morale boost in the United States, demonstrating that Japan could be bombed by US forces. It also caused severe reprisals in China, where the Japanese army punished people in the villages that had aided the US airmen.

The day after his arrival, Seaborg met with Arthur Compton and some of the other people assembling at the Metallurgical Laboratory, which was the code name designed to disguise the bomb-making project in Chicago. He then had lunch at the Quadrangle Club with, among others, his Berkeley colleagues Joe Kennedy and

Art Wahl. They had come to Chicago for a top-secret, two-day conference about the production of plutonium. Though no one knew
whether a chain reaction was possible in the spring of 1942, the scientists and engineers in Chicago were already thinking about how
to build large-scale reactors to produce plutonium—and blenching
at the prospect. The fissioning uranium atoms inside a reactor would
produce intense radioactivity, so the reactor's operators would need
to be shielded by thick layers of concrete and metal. Some of the
elements produced by splitting uranium atoms would be radioactive
gases that could not be retained in the reactor; "a stack 200 to 300
feet high might be used to carry off the gases," Seaborg wrote in
his journal that night. Fission would generate immense quantities of
heat that would need to be removed from the reactor to keep it from
melting or catching fire. And the reactor had to work right the first
time. Once it was operated, its components would become too radioactive for anyone to enter and fix anything that had gone wrong.

Then there was the problem Seaborg had come to Chicago to
solve: What was the best way to separate the plutonium from the
other elements in the irradiated fuel? Several options were in the
running. It might be possible to deposit plutonium from a solution
onto a charged piece of metal. Maybe plutonium could be boiled
away from other elements at temperatures of a couple thousand
degrees. Perhaps plutonium could be added to a mixture of solvents
that later could be separated, with the plutonium attaching itself to
only one of the solvents.

But Seaborg was always partial to the technique he and Wahl had
used in Berkeley. They would dissolve the plutonium in acid and
mix chemicals into the solution that caused the plutonium to form
a solid. They then would separate out this precipitate, redissolve it,
and repeat the process. If this were done over and over, the plutonium should eventually become pure enough to use in a bomb.

The problem, as Seaborg readily acknowledged, was that they
could perform this separation in a laboratory, but could they do it
in a factory? The facilities that would be needed to separate pluto-

nium would be huge, and chemical processes behave differently at large scales. Seaborg and his colleagues had many problems to solve in Chicago, but this was the biggest: Would they be able to scale up their tabletop chemistry by a factor of a billion?

But first Seaborg, as the leader of the Met Lab's chemistry group, had to find enough men to run his laboratory. He began to call and write letters to all the chemists he knew, and then to all the chemists they knew. He couldn't tell anyone what they were doing in Chicago, though he could hint at its importance. "We're working on something that's more important than the discovery of electricity," he would tell potential recruits. "Come here and I'll tell you what it is." Once recruits signed on to the project, Seaborg loved to watch their reactions as he told them about plutonium. "Some stare in disbelief; some are dumbfounded and are glassy-eyed or open-mouthed; others become excited and pour out a torrent of questions." Creating a new element in nuclear reactors to make atomic bombs—it sounded like science fiction.

His group set up in several rooms on the top floor of the university's Jones Hall. They were typical college chemistry laboratories, with metal sinks, concrete benches, and noisy fume hoods. Barricades and a guard station separated the project's chemists from the students and faculty members down the hallway. As the size of the group grew to a couple of dozen men, with an average age of about 25, they fell into a routine. They worked all day, broke for dinner, and then came back to the laboratory after dinner, with Saturday nights and Sunday afternoons off. Seaborg said he was not opposed to romantic activities so long as the night work did not suffer.

Motivation was always near at hand. "It's hard for anyone who didn't live through World War II to imagine the desperation and sense of impending doom that we felt," Seaborg later wrote. In 1942, much of the US Pacific fleet lay at the bottom of Pearl Harbor, Germany controlled continental Europe, and Japan dominated East Asia. Gas and food were rationed throughout the United States, and cities practiced blackouts in case of enemy attack. German science and engineering led the world, and Seaborg and his colleagues were convinced that the Germans were far ahead in building a bomb. "Scientists like me thought less about the benefit of having the bomb than about the potentially disastrous consequences of not having it. Every day, we would follow the war's distant events in the newspapers—German tanks rolling across North Africa, Germans advancing across Russia—events over which we had no control. We could just as easily awaken one day to read the news that Germany had unleashed a powerful new bomb. Every day, every moment, counted."

In the summer of 1942, the leaders of the American bomb project assumed that German scientists had been working on nuclear weapons for more than three years, ever since fission was discovered in Berlin late in 1938. If the Germans had achieved a chain reaction, they could already be manufacturing and separating plutonium. On June 20, Seaborg recorded in his journal Compton's remark that the Germans could have six atomic bombs by the end of the year. "We were fighting for survival, pure and simple."

●

THE SCIENTISTS AND ENGINEERS gathered at the Met Lab in the summer of 1942 had another source of unease. They knew that their work was important to the war effort and that security was essential. They accepted the armed guards in the hallways outside their offices and the badges they had to wear to gain access to their labs. But the military's involvement was going well beyond that.

On June 17, 1942, President Roosevelt approved a plan developed largely by Bush and Conant to have the US Army Corps of Engineers begin building large-scale plants for the separation of uranium-235 and the production of plutonium. Bush had always foreseen turning the project over to the Corps, since it had the expertise to build such plants and a budget big enough to hide the expenditures from Congress. The Corps of Engineers organizes its construction work around engineer districts, and the project's first headquarters was at 270 Broadway in Manhattan. The project was therefore organized under the Manhattan Engineer District, and the effort to build atomic bombs eventually became known as the Manhattan Project.

Ten days after Roosevelt approved Bush's plan to have the army build the production facilities for atomic bombs, Compton held a meeting for the group leaders of the Met Lab. He told them that their job, as outlined by Bush, was to demonstrate the feasibility of a chain reaction and then to design a pilot plant to produce plutonium. However, the Corps of Engineers would be in charge of designing and running the production plants. Furthermore, Compton said, the scientists at the lab might have to be inducted into the army.

The room exploded with objections. They would never be able to do good work as part of the military, the scientists said. They wouldn't be able to meet and talk with each other in the ways that good research demanded. They would have to follow orders and do what they were told, regardless of what they thought they should do.

Already, practically before it had begun, the scientists could feel themselves losing control of the project. They had signed on because of the threat that Germany would get atomic bombs first. They thought that, with a few good engineers and construction teams, they could design, build, and operate the plutonium production facilities themselves. As usual, Szilard, who by this time had moved with Fermi, Anderson, and Zinn to Chicago to work at the Met Lab, was among the most adamant in arguing that scientists should remain in charge of the project: "Those who have originated the work on this terrible weapon," he wrote, "and those who have materially contributed to its development have, before God and the World, the duty to see to it that it should be ready to be used at the proper time and in the proper way."

More bad news was on the way. The Corps of Engineers works by letting contracts to industrial organizations. For the Manhattan Project, the engineering firm Stone & Webster was lined up to build the production plants. The Met Lab scientists would advise Stone & Webster on the design, but the company's personnel, not the scientists, would be in charge of building and operating the plants.

In a famous photograph taken at the University of Chicago, Enrico Fermi and his assistant Herb Anderson are standing in the bottom left and bottom right corners, respectively. Leo Szilard is standing next to Anderson. *Courtesy of Los Alamos National Laboratory.*

To the Met Lab scientists, this was both insulting and dangerous. Stone & Webster had no experience or history with nuclear physics. The project could not possibly work unless it was run by experts. A big engineering firm would have "no knowledge at all of nuclear physics, and very little knowledge of the other engineering problems," recalled Met Lab physicist Eugene Wigner, who even then was working on the design of the production reactors. Furthermore, the involvement of industry suggested the production of atomic bombs on an industrial scale. That made it seem as if the United States would keep building bombs after the war. But only a few bombs would be enough to keep Germany from using any atomic bombs it might manage to build.

Compton remained firm. He told the Met Lab scientists that the construction and operation of plutonium production plants were far beyond their experience and capabilities. They would be much more valuable as advisors than construction managers. That meeting, and subsequent ones on the same issue, ended in an impasse. The controversy would fester throughout the war and beyond.

Chapter 6

PLUTONIUM AT LAST

SHORTLY AFTER COMING TO CHICAGO, SEABORG HAD AN IDEA. AT that point, plutonium had never been made in quantities large enough even to see. Its existence had always been inferred by the radioactive signals it emitted. But the Met Lab desperately needed plutonium to study. How could the chemists, physicists, metallurgists, and engineers learn about its properties without having a sample they could hold in their hands?

In June, Seaborg and his team loaded 300 pounds of a uranium compound into plywood boxes of different sizes and shapes. Trucked to St. Louis, the boxes were wedged into the target area of a cyclotron at Washington University. On July 27 the irradiated uranium arrived back at the University of Chicago. Many of the boxes had cracked open, spilling their highly radioactive contents. Seaborg told the other chemists to wear rubber gloves and lab coats and stay as far away from the irradiated uranium as possible. But whether they were as careful as they should have been is questionable. "We were told to take precautions," wrote Dan Koshland, one of the chemists Seaborg had recruited who later went on to become editor of *Science* magazine. "Almost unanimously, we young scientists discarded this advice because we believed we were in a necessary war against an evil Hitler bent on global domination. With our friends dying on the battlefield, slowing research to be extremely cautious about our own lives seemed inappropriate."

The chemists hauled the irradiated uranium to the fourth floor of Jones Laboratory. In huge cauldrons on the outdoor balconies of the lab, trying to shield themselves behind lead plates as they worked, the chemists dissolved the uranium in ether. They then began the painstaking task of precipitating and dissolving and precipitating and dissolving the uranium, plutonium, and fission products. "We were pretty young, ranging from 20 to probably 30 years old," one of the chemists later told an interviewer:

> The whole operation was carried out in the spirit of what one might say was boisterous fun. At any one time there might be as many as eight or 10 of us shaking up the ether solutions and extracting it. At other times, as during the evaporation, there were of course fewer people involved. There was a lot of kidding and joking. By this time we had come to know each other well, we were all single, we ate lunch together, many of us had dinner together. We might even take a few minutes for a beer together. We were like a close-knit, small family.

Seaborg had hired two chemists who were experts at working with very small quantities of materials, a technique known as micro-chemistry. But these quantities were even smaller than those they had used before, and one of Seaborg's hires coined the term ultrami-crochemistry to describe what they were doing. Watching through microscopes, they used hypodermic needles as pipettes, operated micromanipulators to scale down their movements, and weighed their compounds by hanging them at the end of a single quartz fiber. Seaborg called it weighing invisible material with an invisible balance.

By August 20, they had succeeded in isolating a tiny pinkish speck of plutonium. "Today was the most exciting and thrilling day I have experienced since coming to the Met Lab," Seaborg wrote that night.

It was the first time an artificial element had been created in large enough quantities to see with the naked eye. That evening, at the regular Thursday meeting of the research associates, Seaborg said he "felt like passing out cigars."

○

BY THAT TIME, people had been working with radiation for more than four decades, and its potential for harm was well known. Radiation kills cells by knocking electrons off atoms, and it is particularly tough on fast-dividing blood, skin, and gastrointestinal cells. Pierre and Marie Curie developed radiation burns that took months to heal, and Marie later died of radiation poisoning. Thomas Edison almost lost his eyesight after inventing the first commercial fluoroscope, which used X-rays to image bones, and his fluoroscopy assistant, Clarence Dally, had to have both arms amputated before dying, in 1904, of radiation-induced cancer.

By the 1920s, professional societies representing X-ray workers in the United States and Europe had developed standards to protect against occupational exposures. But even then experts on radiation and health were debating an issue that remains unresolved today. Is exposure to any amount of radioactivity dangerous, or is there a level that the body can tolerate without harm? In the 1920s and 1930s, the latter viewpoint prevailed. Radiation scientists generally assumed that the body could repair damage caused by low levels of radiation. This idea was embodied in the concept of a tolerance dose below which there was no danger.

The work at the Met Lab took this issue to an entirely new level. For the first time, people would be able to manufacture vast quantities of radioactive substances in nuclear reactors. These reactors would produce fission products, plutonium, and other transuranic elements that had never existed before. The people working in reactor facilities would need to be shielded from intense radiation, as would the people responsible for disposing of the reactors' radioactive by-products. Even the people living near reactors would be exposed to unprecedented levels and types of radioactive materials emitted into the air and water; they, too, would need to be protected.

To deal with these and other health issues, Compton set up a health division in the Met Lab that had two main tasks. The first was to protect workers and the public against the radioactivity generated by the project. The second was to learn more about the hazards posed by the radioactive materials. Together, these two endeavors became known as health physics, a name chosen in part to disguise the Met Lab's activities. Health physics remains a thriving discipline today.

To protect workers, the health division adapted standards developed before the war for X-ray technicians and other radiation workers. For example, workers at the Met Lab carried dental X-ray film in their security badges. If the film was fogged when it was developed, a worker had gotten too much exposure.

But detectors had to be used to be effective. Seaborg, for example, was always adamant about protecting the health of his chemists, but they found it awkward to hold a beaker with one hand and a lead shield with the other. Instead, his team rotated people in and out of radioactive areas and drew blood from them every day to see if their white blood cell counts had dropped, which would indicate that their radiation exposures had reached dangerous levels.

Careful monitoring of the Met Lab personnel began to reveal some of the hazards of the atomic age. One time a Met Lab worker was walking by a soda machine when a portable radiation monitor she was wearing went off. When the man delivering syrup to the soda machine had arrived that morning, he discovered that he had forgotten to bring a hose from the truck. Rather than retrieving his own hose, he poked around the lab to find one he could use. Fortunately, the mistake was caught before too many people drank the plutonium-laced sodas the machine had been dispensing.

Learning more about the health effects of radiation required doing research on living organisms. Sometimes the members of the health division volunteered as guinea pigs, exposing themselves to radiation and monitoring the effects. They also did animal experiments, both at the University of Chicago and elsewhere. Once a dog being used in plutonium exposure experiments at the Met Lab escaped into the surrounding neighborhood. The residents of the stately brick mansions on Woodlawn Avenue must have wondered about the white-coated researchers chasing a probably highly radioactive dog up and down the street. Another time the chemists at the Met Lab got nervous about the radioactive wastes they were pouring down the sink. First they spiked wastewater from the lab with a strong banana oil to try to figure out where it was entering the sewer. Failing to distinguish the scent of banana from the other smells in the sewer, they tried unsuccessfully to measure the radioactivity of the sewer water as it flowed by. Finally, they opened the cesspool near the lab that collected solid material and found it to be, in the words of one chemist, "quite radioactive." They closed it up and hoped for the best.

As at other Manhattan Project sites, Met Lab workers who got sick in later years wondered whether exposures to radioactive substances or other toxins might have caused their illnesses. But obvious evidence of ill effects was hard to find. With a few prominent exceptions, scientists who worked on the Manhattan Project had lower mortality rates than the general population. Still, basic questions about the health effects of radiation lingered. These questions would be asked again.

Chapter 7

THE DEMONSTRATION

ENRICO FERMI'S DEMONSTRATION OF A CHAIN REACTION ON DECEMBER 2, 1942, is one of the most famous scientific experiments in history. But it is often misinterpreted—or, rather, the interpretation does not go far enough. Yes, Fermi wanted to demonstrate that a chain reaction using natural uranium ore was possible. But he had a much more immediate goal: he needed to show that the plutonium production reactors being designed by the scientists and engineers at the Met Lab would work.

The end of 1942 was a shaky time for the Manhattan Project. Getting the materials and manpower to build the production plants was going to be difficult as the nation struggled to turn out tanks, aircraft, and guns. The best technique for separating uranium-235 from uranium ore was still unclear. There was no time to build pilot plants; the project would have to move directly from the lab to full-scale manufacturing.

Then, in October, Seaborg had a terrifying thought. Plutonium-239 is a relatively stable isotope, but it occasionally emits alpha particles. If the plutonium in a bomb were contaminated with light elements like beryllium or boron, alpha particles could chip neutrons from their nuclei. These neutrons could set off an atomic bomb prematurely, resulting in a dud. For the Met Lab to succeed in its mission, the plutonium coming from the production plants would have to be almost completely pure.

News of the purity requirements added to the Met Lab's many problems. The scientists complaining about Stone & Webster that summer had been right—the company was not going to be able to build the production reactors. It was already doing too many other things for the Manhattan Project, and it had no expertise that it could apply to producing plutonium. But the army, tightening its grip over the project, wasn't about to turn control over to the scientists. Instead, it started pressuring the firm it wanted to do the job all along—E. I. du Pont de Nemours and Company.

DuPont, which was founded in 1802 in Wilmington, Delaware, as a gunpowder mill and subsequently grew to become one of the largest chemical companies in the United States, was reluctant to get involved. The company was still smarting from the "merchant of death" tag it received after manufacturing munitions during World War I. Like Stone & Webster, it was involved with other World War II projects and did not want to spread itself too thin. When DuPont officials heard how pure the reactor-produced plutonium would have to be, they were even more wary.

The army turned up the pressure. It said that the plutonium project needed one of the country's best chemical companies if it was going to succeed. The outcome of the war might hinge on DuPont's decision. No other firm could do the job. Company officials were torn. They didn't want to take on the project, but they had a hard time rejecting the army's appeals.

When in doubt, form a committee—and one of the most important committees of the entire Manhattan Project took shape that fall. Its job was to examine not only the Met Lab but also the isotope separation projects going on around the country to decide which parts of the project should be emphasized and which deemphasized or canceled. The chair of the committee was Warren Lewis, a chemical engineering professor at MIT who had served on the National

Academy of Sciences committee the previous year. A second member was Eger Murphree, though he fell sick in November and did not participate in the committee's deliberations. The other three members were all DuPont engineers, named to the committee in the hope that their review of the plutonium project would help convince the company to take it on.

The most influential of the three was Crawford Greenewalt, just 40 years old but already a rising star in DuPont. A lanky New Englander who carried himself with patrician grace, he had earned a bachelor's degree in chemical engineering from MIT, went to work for DuPont, married the sister of DuPont president Irénée du Pont Jr., and played a key role in the development of nylon. Greenewalt would later become president of DuPont himself. His work on the Manhattan Project was both a pathway and rite of passage to that position.

The committee traveled first to Columbia University to inspect the isotope separation work being done there. The technique, which relied on forcing gaseous uranium through a fine wire mesh, appeared likely to work eventually. But isotope separation was going to be a long and arduous process. If isotope separation was the only route to a bomb, success was far from guaranteed.

The committee then took the train to Chicago, arriving on Thanksgiving morning, November 26. Gathering in Eckhart Hall's conference room over platters of turkey, the committee heard from Compton first. He said that the Met Lab could produce enough plutonium for a bomb by 1944 and achieve regular production in 1945. Compton listed seven problems that had to be solved for plutonium production to work. Progress was being made on them all, he said.

Then it was Seaborg's turn. Could he produce enough plutonium, the committee asked, of the purity required, for a successful nuclear explosion? He said that he could, so long as he had enough good chemists and metallurgists to do the job. The committee was impressed but noncommittal. That evening, still undecided on whether to recommend that the plutonium production effort continue, the committee members boarded the train for California to review the isotope separation work being done at Berkeley.

•

BY THE DAY OF Fermi's experiment, the review committee had returned to Chicago from California. It was Chicago's coldest December 2 of the previous half-century—just below zero Fahrenheit, with a howling wind to compound the misery. Over the previous three weeks, Fermi and a rag-tag team of physicists, technicians, and young men hired from the rough neighborhood north of campus had built a massive pile of graphite blocks and embedded uranium spheres in a racquets court beneath the west stands of the university's abandoned football field.* The pile was controlled by wooden poles covered by strips of cadmium metal, which is a voracious neutron absorber. So long as these poles were in the pile, the chain reaction could not occur. But as the poles were withdrawn, the flux of neutrons within the pile would increase, to the point, Fermi hoped, that the pile would be producing more neutrons than it was absorbing.

With his colleagues crowding the balcony of the racquets court, Fermi, at 9:45, began ordering that the control rods be withdrawn. Fermi knew the significance of what he was doing, and he had planned his performance accordingly. Each time the main control rod was pulled out another foot or six inches, he took his slide rule from a pocket, made a few calculations, and jotted some numbers on its reverse side. "This is not it," he would say, pointing to the chart recorder that was documenting the neutron flux. "The trace will go to this point and level off."

About midmorning, Compton got a call in the conference room

---

* It was not a squash court, as has been reported in the past. A squash court, even a doubles squash court, would not have been large enough for Fermi's pile. Plus, courts built for racquets, which is a game invented in the 18th century that preceded squash, have a balcony on one end where the scorekeeper could sit, and Fermi put his electronic equipment on this balcony.

where he was again meeting with the review committee. He should come to the west stands to watch the experiment, he was told. But he could bring along only one member of the committee, because the balcony was crowded. Compton chose Greenewalt, explaining that because he was the youngest member of the committee he "would probably remember longer than the others what he would see, and it should be something worth remembering." Greenewalt later ventured that perhaps he was the most expendable.

The crucial moment did not come until 3:20 that afternoon. "This is going to do it," Fermi told Compton after directing that the control rod be withdrawn one more foot. "Now it will become self-sustaining. The trace will climb and continue to climb. It will not level off." As the scientists on the balcony crowded around Fermi, the needle of the chart recorder began to rise. In the pile, neutrons were splitting uranium atoms, which released neutrons that split more uranium atoms, which repeated the process again and again. Leo Szilard's vision of the chain reaction had been realized. The atomic age had begun.

After Fermi powered down the reactor, Compton and Greenewalt walked back across campus toward the conference room. Greenewalt's "eyes were aglow," Compton later wrote. "His mind was swarming with ideas of how atomic energy could mean great things in the practical lives of men and women. . . . Here was a source of endless power that could warm peoples' homes, light their lamps, and turn the wheels of industry." Greenewalt spun his reaction differently: "What I was really thinking about was how happy I would be to get home."

The other members of the review committee could tell as soon as Compton and Greenewalt walked through the door that the experiment had been a success. Two weeks later, DuPont agreed to take on the job of building and operating the plutonium production plants.

Seaborg was not among the people on the balcony on December 2, but word spread quickly through the Met Lab that afternoon. That evening, he wrote in his journal, "Of course we have no way of knowing if this is the first time a sustained chain reaction has

Because Chicago Pile 1 was secret, no photographs were taken of the completed structure. Based on interviews with eyewitnesses and published accounts, *Chicago Tribune* staff artist Gary Sheahan painted this rendition of the observers crowded around Fermi on the balcony of the racquets court. *Courtesy of the US National Archives and Records Administration.*

been achieved. The Germans may have beaten us to it. . . . [But] one thing is certain: though Fermi has demonstrated that we now have a means of manufacturing [plutonium] in copious amounts, it is the responsibility of chemists to show that [plutonium] can be extracted and purified to a degree required for a working bomb."

THINGS HAPPENED QUICKLY after December 2. A week later, the federal committee overseeing the Manhattan Project decided to move ahead full speed with both the isotope separation facilities and the plutonium production plants. The initial idea had been to place the plutonium plants in Tennessee, near the immense factories that would be built to isolate uranium-235. But the production reactors would compete with the separation facilities for electricity. Plus, they were too dangerous. A reactor accident could shower Knoxville and other downwind communities with radioactivity. The plutonium facilities would need to be built elsewhere.

Meanwhile, as soon as DuPont agreed to take on the plutonium project, the company convened a group of top engineers and executives to oversee the company's involvement. To avoid perceptions of war profiteering, DuPont asked for a contract that would cover its costs plus a fixed fee of one dollar. The contract with the company called for any patents arising from the project to belong solely to the federal government. It also indemnified DuPont from any liabilities the company might incur.

In the middle of December, Corps of Engineers Colonel Franklin Matthias attended a meeting in Wilmington at which DuPont engineers and Compton established the criteria a plutonium production site must meet. Back in Washington, DC, Matthias told his commanding officer that he needed to reread Buck Rogers cartoons to get up to speed on the project. A few days later, he and two DuPont engineers departed for the West Coast to scout out potential locations. One was especially promising: a broad, sparsely populated plain in south-central Washington State near the small farming town of Hanford.

PART 2

A FACTORY IN THE DESERT

"Some of the flattest, most lonesome territory I had ever seen."   —Glenn Seaborg

Chapter 8

THE EVICTED

ON MARCH 6, 1943, JUST AS THE FIRST WARM BREEZES FROM THE
south had begun to hint of spring, Frank and Jeanie Wheeler of
White Bluffs, Washington, received the following letter:

> Dear Sir or Madam:
> Re: U.S. vs. Alberts, No. 128
> You are advised that on February 23, 1943 the United States of
> America instituted the above proceeding to acquire certain real
> property by condemnation, including lands apparently owned or
> occupied by you. Pursuant to the court order that day entered in
> said proceeding I herewith enclose a copy of such order, which will
> serve to advise you that the United States was on said date given
> the right immediately to take possession of such property. . . .
> Respectfully,
> Hart Snyder
> Special Attorney for the Department of Justice

The Wheelers stared at the letter in disbelief. The attached legal
notice said that the land was being taken "for military, naval or other
war purposes." What could the government possibly want with their
orchards and fields? It was far away from any military installation or
transportation center and could be farmed only with constant effort.
That part of Washington State was scorching hot in summer and
cold and foggy in winter. It didn't make sense.

By 1943 the Wheelers had been living in eastern Washington for more than three decades. They had raised six children there, four of whom had gone to Reed College in Portland and one of whom became a Rhodes Scholar. "It was a wonderful place for the children," Jeanie recalled many years later. When the Columbia flooded in spring, the older kids rowed onto the river to capture wood floating downriver from distant forests, which would supply the family with firewood for the year. In the summers, they could swim in the river or in the irrigation canals that crisscrossed the desert.* On warm summer evenings, as heat lightning flashed in the purpling sky, Frank and Jeanie could sit outside and watch the Columbia roll by in front of the terraced backdrop of sand, gravel, and clay that had given the town of White Bluffs its name. The Wheelers grew apples, peaches, apricots, and pears in their orchard, which they shipped to markets from the nearby rail station. Agricultural prices were low for much of the 1920s and 1930s, and many of the farmers in the area had left. But prices for food were going up with the United States at war, and the Wheelers were looking forward to a profitable year.

The Wheelers soon learned that everyone in White Bluffs had gotten the same letter. So had the 300 or so residents of the nearby town of Hanford, named after a Seattle lawyer who helped bankroll a nearby irrigation district. Everyone in Richland, a town of about the same size 30 miles south of White Bluffs, also had gotten the letter. Along with the people who lived on the surrounding farms, about 1,500 people had to leave their homes. "It was a terrible shock," said a resident of Hanford. "The only thing that made it credible to us was because of the war. Our town had been chosen for the war effort. We were so patriotic. Although we could go along with that idea, it was still a terrible blow." Even the dead had to leave. The army dug up 177 graves at the White Bluffs cemetery and moved the remains to the town of Prosser on the other side of Rattlesnake Mountain.

---

* Technically, the land around Hanford is shrub-steppe, not desert, but even people who grew up there think of themselves as desert dwellers.

In May, appraisers arrived at the Wheelers' farm. "What's the old barn over there?" one asked of the almost completed house Frank and Jeanie's son had been building. Another said, "You should be glad to get out of this godforsaken place." When their appraisal arrived, the Wheelers were shocked anew. Their 40 acres of property, with three houses, barns, wells, and 20 acres of thriving orchards, were valued by the government at $1,500. Their neighbors were equally appalled. "They appraised my father's 30 acres at $1,700," said one. "The pump and well alone had cost us $1,900."

Frank wrote to the children who had moved elsewhere. "If you want to see this place again, you had better get over here." Only one could get away from her wartime job. Frank and Jeanie's daughter Helen arrived from Seattle just as a bulldozer was ripping out the family orchard. "I've never gotten over that scene," she later said. "I many times wished I had not come."

Finally, the day to leave arrived. "They came that morning to take us away so early that we just finished breakfast and the stove was still hot," said Jeanie. The Wheelers moved in with another daughter in Seattle and eventually retired to an island in Puget Sound. "We always thought that they should have left us there," said Jeanie many years later, "that peaches were a better crop than atom bombs."

FRANKLIN MATTHIAS HAD BEEN BUSY since he'd flown over Richland, Hanford, and White Bluffs the previous December and had decided that the area would be perfect to build the plutonium production facilities. His boss, General Leslie Groves, had traveled to the site in January 1943 and had agreed with Matthias about its merits. Back in his Washington, DC, office, Groves arranged for Secretary of War Henry Stimson to seize the land under the War Powers Act. Eventually the federal government acquired more than 600 square miles of property, an area 10 times that of Washington, DC.

Groves still had not announced who would build and operate Hanford when he called Matthias into his office in mid-February.

The town of White Bluffs was named after the terraced cliffs on the left bank of the Columbia River. The Saddle Mountains rise in the distance. *Courtesy of Jim Stoffels.*

"I have a promise from the Chief of Engineers that I can have anybody that you want in the Corps of Engineers who's not on combat duty," he told Matthias. "I wish you'd review the possibilities and recommend somebody to me."

Matthias nodded and turned to leave. Just as he was opening the door, Groves called to him "By the way, if you don't find somebody I like, you're going to have to take over that project."

Matthias hesitated for only a moment. He turned to Groves and said, "General, there isn't anybody I can recommend."

"All right," said Groves, "you're it."

Matthias was hardly the obvious choice. Just 34 years old in the spring of 1943, he had joined the Reserve Officers' Training Corps as an undergraduate at the University of Wisconsin-Madison, but he did not think of himself as a military man. He served in the Wisconsin National Guard after college and then went into private industry, working on dams with the Tennessee Valley Authority and designing part of the Delaware Aqueduct. By the spring of 1941 he could tell he was going to be recalled to active duty, so he opted to enlist rather than being recalled. He thought he would serve for a year and then be done with it, but the attack on Pearl Harbor scuttled those plans. Given his background, he was assigned to work with the Corps of Engineers, where he caught the eye of General Groves while they both were working on the construction of the Pentagon. Six months later, he was in charge of building Hanford.

Matthias had four huge tasks to complete. The first was to acquire the supplies and equipment he needed and have them shipped to the site. Groves had arranged for the Manhattan Project to receive the highest possible priority rating for materials acquisition, but Matthias struggled throughout the war to explain why he needed such expensive supplies for a project that he was not allowed to describe. A big boost was the completion of the Alaska Highway in 1942. For the next few years, massive construction equipment from Alaska rattled down the Columbia River rail line to Hanford.

Second, he had to hire and house the thousands of construction workers who would build the plants. The soon-to-be-abandoned town of Hanford looked to be a good spot for a construction camp. It was on a broad plain next to the river and had been optimistically platted by its founders to be larger than Chicago. The site could house the barracks and trailers for the construction workers and then be abandoned when the startup of the reactors posed risks to anyone living nearby.

Third, he had to build a permanent town to accommodate the engineers and operators who would build and run the plants. Here Richland, on the southern end of the land acquired by the government, was the obvious choice. It could be separated from the reactors by a safety buffer of more than 30 miles. And Hanford workers who did not rank high enough to live in Richland could live in the towns of Kennewick and Pasco a few miles farther down the Columbia.

Finally, Matthias had to combine all these pieces to get the facilities up and running in time to make a difference in the war. When he set up his initial office at the site in February 1943, the production plants were still being designed. But the outlines of Hanford Engineer Works, as it would become known, were beginning to take shape. The reactors could go in what was labeled the 100 area at the far northern end of the site, along the right bank of the Columbia River. South of the reactors, on a broad plateau topped by a sharp ridge of basalt known as Gable Mountain, the plants for separating plutonium from the irradiated uranium could be built—this would be the 200 area. Finally, the safer facilities for producing the uranium fuel and conducting tests—the 300 area—could be just north of Richland, closer to the employees' homes.

It was the largest construction project ever undertaken in Washington State, larger even than the construction of Grand Coulee Dam the decade before. Yet Matthias had to keep the entire operation as secret as possible. On February 26, 1943, he walked into the offices of the *Pasco Herald* and asked to talk with the editor, Hill Williams. He told Williams that a huge plant was going to be built near Richland but that it had to remain absolutely secret. He wasn't picking just on the *Herald*, he said—he was making the same request of all the other newspapers and radio stations in the Pacific Northwest. Don't worry, he told Williams, as soon as the story breaks, you'll be the first to know.

That same day Matthias talked with the editor of the *Walla Walla Union-Bulletin*, and a few days later he visited the *Yakima Morning Herald*. Matthias and other officers talked to editors at the *Kennewick Courier-Reporter*, the *Prosser Record-Bulletin*, and the major papers in Seattle, Portland, and Spokane. All agreed to his request that they ignore a story that normally would have warranted banner headlines.

The agreement was not foolproof. An article in the March 2 *Seattle Times* said that hundreds of displaced people were "protesting that they were getting 'worse treatment than the Japs,'" many of whom were being forcibly relocated from western cities into internment camps. An April 1943 article on the land seizures in an Idaho newspaper led Matthias to lament that "trying to restrict publicity on this project is like keeping water in a sieve."

Still, the secrecy campaign was surprisingly effective. For the next three-and-a-half years, the word *Hanford* largely disappeared from the newspapers of the Pacific Northwest.

◉

THIS WAS NOT THE FIRST TIME the US government had upended the lives of people living along that stretch of the Columbia River. On October 16, 1805, Meriwether Lewis, William Clark, and the other members of the Corps of Discovery emerged from a difficult section of the Snake River to discover several hundred Indians camped at the river's junction with the Columbia. After hauling out their canoes and making camp, they were welcomed by a delegation of about 200 Indians, who formed a half circle around the men and "Sung for Some time," as Clark wrote in his journal. The next day, Clark and two men paddled 10 miles up the Columbia to the present-day site of Richland, passing lodges made of driftwood, reed mats, and furs. "This river is remarkably Clear," Clark wrote. "I observe in assending great numbers of Salmon *dead* on the Shores, floating on the water and in the Bottoms." To the north and east of the river stretched a plain backdropped by distant ridges with "no wood to be Seen in any direction"—the future site of Hanford.

The tribe that occupied the area upstream of Richland was known as the Wanapum, or water people. They called the Columbia *Ci Wana*, which meant big water. A seminomadic tribe, they traveled west in the summer to pick berries and hunt in the mountains. In the fall, they returned to the Columbia to fish, preserve the salmon they caught, and manufacture tools. Along the riverbanks the Wanapum had pit-house villages, fishing sites, trading camps, quarries, food caches, and cemeteries. They traded, socialized, and intermarried with the many other Native groups that gathered at the intersection of the Yakima, Snake, Columbia, and Walla Walla rivers. Their young men journeyed to the nearby heights on spirit quests.

The arrival of Lewis and Clark marked the first time white men had visited that part of the Columbia River, but the influence of whites had already been felt. Diseases introduced to the coastal tribes by traders and settlers had been decimating the inland tribes for several decades. From a population of almost 2,000 in 1780, the Wanapum declined in numbers to just 300 in 1870. By that time, members of Indian tribes throughout the Northwest had signed treaties forfeiting their claims to the land in exchange for out-of-the-way reservations. But the Wanapum never signed a treaty. They continued to travel from place to place, doing temporary work for hire, picking berries in the summer, and fishing in their accustomed spots on the Columbia. During the middle of the 19th century, as the tribe continued to decline, a Wanapum shaman named Smohalla began to preach a doctrine known as the Washat religion. If his people would shun the white man's ways and live as their ancestors had done, the dead ancestors of the Wanapum would return to life, and the world would be restored to how it was before the arrival of white men. Smohalla's teachings spread widely among the inland tribes of the Pacific Northwest, contributing to the Ghost Dance religions that spread among the Plains Indians beginning in the 1870s.

By the time Matthias arrived at the Hanford nuclear reservation in 1943, only about 30 Wanapum continued to occupy their winter camp on the Columbia. Mattias could logically have ejected them from the site, just as he had the white settlers. But Matthias had an unexpected affinity for the Wanapum. When they asked for access to their customary fishing grounds in the middle of the secret federal reservation, Matthias agreed, not even requiring that they undergo a security clearance. He provided trucks to bring the Wanapum onto the nuclear reservation and back to their camps each day. The Wanapum "insist on maintaining their independence . . . to fish on the Columbia River," Matthias wrote in his diary. "I do not believe that their loyalty can be questioned." After the war the federal government did bar the Wanapum from Hanford, sealing off their graves and cultural sites behind barbed wire fences. But during World War II at least, they were not totally excluded from their homelands.

Chapter 9

THE BUILDERS

BILL PORTER KNEW HE SHOULD WORRY WHEN HE GOT OFF THE NORTHERN
Pacific train in Pasco after a long ride from Tennessee. A group of
workers was waiting to leave. "Suckers, suckers, suckers," they
chanted at the new arrivals.

In the spring of 1943, newspaper advertisements and recruiting
pamphlets began to appear in cities and towns across the country. "War Construction Project," read one such advertisement.
"NEEDED by E. I. du Pont de Nemours & Company for PACIFIC
NORTHWEST":

Laborers
Millwrights
Carpenters
Sheetmetal men
Reinforcing ironworkers
Iron worker welders
Structural iron workers
Auto mechanics
Auto oilers
Heavy equipment mechanics
Machinists
Patrolmen
Protective firemen

Registered nurses

Physicians

"Living Facilities Available for all Persons Employed," the ad stated. "Attractive Scale of Wages. Work week 54 hours. Time and one-half for work in excess of 40 hours."

Matthias needed tens of thousands of workers to build the facilities that the DuPont company was planning for Hanford. But many young, able-bodied men were at war, and Matthias was competing with other wartime construction projects across the country. Eventually, DuPont recruiters interviewed more than a quarter-million workers for jobs at Hanford and hired almost a hundred thousand. But many failed to arrive, and others left within a few days or weeks after they got a look at what the job entailed. With all the churn, the peak workforce was about 45,000 in May 1944.

Known as "boomers" in World War II—the name given to men who moved around the country from one boom town construction project to another—they were a motley crew. Many of the male employees were over the maximum draft age of 37 years. Of the younger men, many were physically unfit for military service. Recruitment standards were low. Matthias once asked a recruiter, "Is it true that you people would hire a man as a carpenter if he could just identify a hammer?" The recruiter laughed. "No, we're not quite that tough. If the man can convince us that his father would have known what a hammer was, we take him."

Many newcomers got their first surprise when the sun came up after their nighttime arrival on the train. Washington is a state divided by a common mountain range. The Cascade Mountains that roughly bisect the state from north to south squeeze the moisture out of the clouds blowing off the Pacific Ocean. The western half of the state, where most of the big cities are located, is green and sodden. The eastern half, where Hanford is located, is drier than west Texas. "I had thought there was a lumberjack behind every tree and salmon jumping out of the Columbia River," said Claribel Chapman, who was living in Florida when she heard about Hanford and decided to move. "I had no idea there was a desert in Washington."

The next surprise was the weather. The basin in which Hanford sits is low and relatively warm in winter. But when the wind shifts into the northeast and arctic air blows from Canada, temperatures can drop below zero. In the summer, beneath a blistering sun, it can be well above 100. And the temperature is only the beginning. The contrast between wet and dry in Washington State can create fierce winds that roll down the eastern slopes of the Cascades like a runaway freight train. When the wind blew hard, dust storms the equal of anything in the Dust Bowl descended on Hanford. The workers called them termination winds, because as soon as the wind quieted, workers lined up to collect their termination checks and get out of town.

By the summer of 1943, a strange alternative version of the town Hanford's founders had envisioned was beginning to rise from the desert. Gigantic H-shaped barracks, each with beds for 190 people, spread like mushrooms across the tableland next to the river. Hundreds of hutments held 10 to 20 men each, and a nearby trailer camp had 3,600 lots. A ramshackle city, never meant to last more than a few years, began to take shape. It had banks, a hospital, power plants, barbershops, fire and police departments, a grocery store, a post office, baseball fields, and a swimming pool. Streets in the trailer camp were named after ongoing battles in the Pacific. Streets in the main camp had the names of military leaders—Eisenhower, Doolittle, Patton. It was the largest construction camp ever assembled in the United States and, briefly, the fourth largest city in Washington State.

The construction workers put in nine-hour days Monday through Saturday, with Sunday work added as needed. Managers worked from 10 to 20 hours a day and rarely got Sundays off; some just stayed on the job permanently, especially since there was so little to do otherwise. Pay was about $1.00 an hour for laborers up to $1.85 for skilled workers such as plumbers, steamfitters, electricians, and

The Hanford construction camp on the banks of the Columbia River. *Courtesy of the US National Archives and Records Administration / US Department of Energy.*

bricklayers. The specialized welders known as leadburners made the most—$2 an hour.

Despite severe manpower shortages, Matthias initially resisted hiring African Americans, even though an executive order signed by President Roosevelt in 1941 required him to do so. Few African Americans had lived in the area previously, and most of them were working for the Northern Pacific Railroad. Matthias also assumed that he would have to build separate housing if he recruited Black workers, which would add to his administrative tasks. It seemed easier to hire only whites.

But soon the pressure to hire African American workers became too great, imposed both by Black leaders in the Northwest and by the need to get the job done. By the end of the war, about 15,000 African Americans had arrived to work at Hanford. Many lived in barracks in the construction camp designated for Black workers. Most of the others lived in a ramshackle part of Pasco. Though many moved away after the construction was done, about 1,000 African Americans still lived in Pasco in 1950.

Hanford also had jobs for women, both white and Black, who made up about 10 percent of workers at the 1944 employment peak. They served as cooks, servers, secretaries, clerks, nurses, teachers, typists, and laboratory assistants. They lived in separate barracks, with a barbed wire fence and guards to keep the men out. To provide the women in the barracks with amenities that would keep them from leaving, Matthias hired a former dean of women at Oregon State University, Buena Maris, who tried to make the women's lives as bearable as possible. She organized a library, sports activities, parties, and shopping trips to Pasco. Each barracks had a housemother who kept things running smoothly and tried to solve any problems that arose. When Matthias once asked Maris about the women's highest priority, she said that the camp's gravel walkways were ruining their tightly rationed shoes. The next morning, Maris was amazed to see the walkways being paved with asphalt.

A DuPont booklet for incoming women employees entitled "Dear Anne—a letter telling you all about life in Hanford" tried to counter the bad with the good:

> It isn't an "easy life," but it isn't too hard a one, either. You won't be "roughing it," for a very sincere attempt has been made to care for the everyday needs of the personnel here.
>
> Come prepared to do your best, and to be a good sport about things. You'll have the satisfaction of knowing you're doing your part to bring Victory sooner, in a war job that is vital to our country.

For the men, as one resident said, Hanford Camp "was like Saturday night every day." Card and dice games spread throughout the barracks, while prostitution centered on the scrubby bushes down by the Columbia. The chief of police noted that workers "hit the booze very hard." Closing time at the bars in the construction camp was 11, so early that many patrons did not want to leave. Windows were covered by metal bars so that the police could throw teargas inside to vacate the revelers. Intoxication was by far the most common cause of arrest but rarely led to dismissal given the need for workers. Drunks dried out overnight and went back to work the next day.

To keep employees from leaving, Matthias tried to keep them entertained. The craft unions organized baseball teams—which were integrated, despite the lack of integration in most of the other social activities—and attendance at ball games approached that at big league games. When workers were told that they could build a hall for events, they took just three weeks to erect an auditorium that could accommodate 3,000 dancing couples. Mondays featured boxing or wrestling. Bands came to play on Thursdays, Fridays, and Saturdays, including national acts—Kay Kyser, Tiny Hill, Ted Weems, Jan Garber, and others.

Many of the construction workers chafed at the primitive living conditions. Still, many remember the experience as a great adventure and as their own contribution to winning the war. "It was exciting," said one many years later. "I had three hots and a cot. I had a good-paying job that wasn't too hard. I was free to come and go as I pleased, and nobody was shooting at me." They were patriotic about what they were doing, even though they had no idea what the gigantic plants they were building would make. In 1944, the craft unions organized a campaign to ask everyone for a day's pay, raising $162,000 in seven weeks. With the funds, the unions bought a four-engine B-17 bomber for the US Army Air Forces. Named "Day's Pay," the bomber flew from Boeing Field in Seattle to the Hanford airstrip to be presented to the Fourth Air Force. "This activity, conceived by the workmen and handled by them, . . . was the most effective single morale booster during the job," Matthias recalled. It did more "to develop an attitude of teamwork and desire to help the war than any other thing."

Matthias sometimes used workers' patriotism to sway labor disputes. Once he was fishing with a friend off the mouth of the Columbia River when a patrol boat came alongside and told him that he needed to call Hanford—the pipefitters were threatening to go on strike the next morning. Matthias drove all night and the next morning was on the auditorium stage in front of 500 union workers. "We have a contract with you that you do not strike," he argued. "There must be some people that are leading you into this, and they're wrong and they're against us, and I'd like to have them all be picked up and sent back to Germany where they belong." At this point, given the crowd's antagonism, Matthias thought he was going to be shot if one of his listeners was armed. "Look, take it easy," he continued. "I'm not calling you traitors, but some of you are acting like it. Now how about going back to work and doing what you promised, and what we need badly?" The reworded appeal worked. The men dispersed and the strike was canceled.

Harry Petcher and his wife Maxine arrived at Hanford on July 3, 1943, just as the construction camp was taking shape. Ineligible for the draft because of flat feet, Petcher had been managing a nightclub in Chicago while Maxine worked as a cocktail waitress. "We got off that train and I looked around and I said, 'Honey, I don't know if we're going to make this or not.' She was a pretty sturdy girl, and she said, 'No, we're going to do it.'" The next day it was 105 degrees as they watched a seven-horse parade commemorate Independence Day. By the end of the week, Maxine had a job as a mess hall supervisor; Harry went to work as a butcher.

Hanford camp eventually had eight football field–sized mess halls that could each serve 2,700 people three times a day. Tables were set with silverware and napkins, after which they were covered by butcher paper to keep the sand off until right before the meal was ready. Workers sat at long tables and served themselves from bowls and platters. When a pot of coffee or platter of fried chicken was empty, someone held the container in the air and a server would rush to replace it with a full one. Men ate quickly and banged their hands on the table when they were waiting to be served. "It looked like a prison camp to me," Petcher recalled.

Once, one of Petcher's subordinates put sugar instead of salt in the meatballs. "Everything was fine until we got a police call," said

Petcher. "There was a riot in the mess hall. Everybody was standing on their benches, picking these meatballs up and throwing them at the cooks. Getting hit by one of them was like getting hit by a golf ball. We had to call the riot squad."

Eventually, Petcher was put in charge of the box lunch department, which at its peak was making more than 50,000 box lunches a day. The lunches contained sandwiches, fresh fruit, sometimes a salad or cold baked potato, and, in the summer, a couple of salt tablets. A big problem for the cooks was spreading mayonnaise on the sandwich bread, which slowed down the entire operation. The husband of one of Petcher's employees solved the problem by putting electric heaters in a paint gun and spraying mayonnaise onto the bread.

Eventually, the Petchers moved to a ranch outside the small town of Connell about 30 miles northeast of Hanford. They had friends over to the ranch for barbecues. They bought horses from the Wanapum and ate fruit from the orchard on their property. "I hate to say I was a rancher, because I'm not, I'm a nice boy from Chicago," Petcher recalled many years later, when he was retired and living outside Seattle. Still, he remembered, "the life was good."

○

HANFORD CAMP WAS BRASH and unapologetic. When problems arose, its occupants moved quickly to solve them. The workers came there to do a job, and they worked hard until the job was done. They didn't ask what they were doing or question whether something was necessary. In all these ways the camp and its residents reflected the attributes of the man who was responsible not only for Hanford but for the entire Manhattan Project, the only other man, besides Enrico Fermi and Glenn Seaborg, without whom nuclear weapons almost certainly would not have been ready by the end of World War II: Leslie Richard Groves.

The son of a restless army chaplain, Leslie Groves—Dick to his friends—spent the first part of his life, from 1896 to 1901, in army housing in Vancouver, Washington, just across the Columbia River from Portland. His father, a strict and austere man, was away for much of that time, in Cuba, the Philippines, and China. He returned to the United States when Dick was five and quickly imposed his rules on the household. After church on Sundays, Dick and his two brothers had to stay inside to read religious books. In the evenings, Chaplain Groves and his sons studied *The World Almanac* so rigorously that all three boys could name not only all the presidents and vice presidents but also the populations of the country's hundred largest cities. Chaplain Groves and his wife pitted their sons against each other and against the world. "Second best in class is good," Groves's mother once wrote to his brother. "Next year we will try hard for first." Years later, Groves's own son, who was raised the same way as his father, recalled, "If it is a game, you win it. If there is a class, you stand number one."

While Groves was in the first through tenth grades, his family moved six times, ending up finally in Altadena, California, with summers spent in Fort Apache, Arizona, where Chaplain Groves was stationed to recuperate from the tuberculosis he had contracted overseas. The summer he turned 15, Groves spent the summer alone with his father at Fort Apache while the rest of the family stayed in California. On the days when his father worked at the fort, Groves played tennis at the Indian Agency, often not leaving the courts until after dark. He came to love the game and was a formidable player his whole life, even when he was carrying some extra pounds, which was most of the time. On his father's days off, they rode horseback into the surrounding scrubland and wooded hills, fished for trout, and often just sat in the shade of a tree and talked. Groves always admired his father, despite his severity and long absences from the family. His father taught Groves to excel in whatever he undertook, and Groves never let his father down.

Halfway through Groves's junior year in high school, his father's regiment moved to Fort Lawton on a wooded point of land extending into Puget Sound just north of downtown Seattle, and Groves entered nearby Queen Anne High School in January 1913. By this time, he had decided, under the influence of a family friend and his father's military heritage, that he wanted to attend the US Military Academy at West Point, even though his parents disapproved of the idea and he had not previously been much of a student. But a traumatic event the summer after his junior year changed Groves's approach to life. His mother, at the age of 47, died suddenly of a cerebral hemorrhage. Chaplain Groves was away again and could not get back in time to bury his wife, so his son handled the funeral. "Dick has been a wonder of thoughtfulness and ability," his aunt wrote in a letter, "and has tried to save all of us from anything harrowing." For the rest of his life, his mother's portrait hung on the wall above Groves's dresser.

That fall, Groves enrolled at the University of Washington while simultaneously completing his senior year at Queen Anne High School. He worked incredibly hard and earned a high school diploma with his classmates, but his scores on the West Point entrance examination were not high enough for a presidential appointment. Groves was undeterred. He won acceptance to the Massachusetts Institute of Technology and, in the summer of 1914, endured a long and sweltering trip to the East Coast on the Canadian Pacific Railway, which was filled with Englishmen trying to get home to fight in World War I. Living in a fourth-floor walkup near Copley Square, he allotted himself $3.50 per week for food, with which he could buy two hot dogs and two slices of bread with mayonnaise for lunch, if he skipped breakfast, and a 35-cent dinner at the student dining hall.

The next time he took the entrance exam for West Point, during the spring of his sophomore year at MIT, he was much better prepared. This time his scores were exemplary, and on July 15, 1916, he trudged up the hill from the train station on the western shore of the Hudson River to the West Point campus. As he later wrote to his grandchildren:

Entering West Point fulfilled my greatest ambition. I had been brought up in the army, and in the main had lived on army posts

all my life. I was deeply impressed with the character and out-
standing devotion of the officers I knew. I had also found the
enlisted men to be good solid Americans and, in general far supe-
rior to men of equal education in civil life.

At West Point, Groves entered a world of regimentation, discipline, and
competition—and he thrived in it. Though his grades had not been
particularly good at the University of Washington or MIT, he excelled
academically at the Academy. Ranked twenty-third in his class during
his first year, he ranked second the following year. He did not socialize
much and had few close friends. The evenings he devoted to studying
and writing letters to his family.

Groves's time at West Point was cut short by America's entry into
World War I. His class graduated on November 1, 1918, more than
a year and a half early, to meet the demand for wartime officers. Ten
days later the armistice ended the war. As was the case for almost
all the top-ranking cadets at the Academy, Groves had chosen to
join the prestigious Corps of Engineers. Now he found himself in a
peacetime army swollen with young officers looking to advance.

Groves spent the next 22 years alternating between the educational institutions that prepare aspiring young officers for high
command and postings where he could hone and display his skills.
Between his stints at school, he built camps and a railway in Georgia,
oversaw the construction of a mountainous military trail on Oahu,
improved shipping channels in Texas, helped survey a possible canal
route across Nicaragua, and supervised the building of the massive
Fort Peck Dam in Montana. During a stint at Fort Worden on the
Olympic Peninsula across Puget Sound from Seattle, he wooed and
married Grace Wilson, whom he had met on an army base when he
was 15 and she was 14. Over the next decade, they had two children
and traveled even more frenetically than Groves had as a boy.

Groves had an odd combination of characteristics. He was smart,
organized, and ambitious. But he also could be fantastically rude,
arrogant, and condescending. As Matthias said, "He was really a

genius, but he didn't spend much time trying to make people like him." He never raised his voice, but he openly and harshly berated people with whom he disagreed or was disappointed. He struggled with his weight throughout his life and often had candy and chocolate nearby except when he was on one of his 1,000-calorie-a-day diets. He was targeted for leadership and rose more rapidly through the ranks than any of his West Point classmates. But he made many enemies along the way who worked to undermine his successes. A colleague recalled, "No one took this man lightly. I was careful to have no trouble whatsoever with him. When you looked at Captain Groves, a little alarm bell rang 'Caution' in your brain."

Most profiles of Groves mention his role in the construction of the Pentagon right before he took on the leadership of the Manhattan Project, but an even more important experience came right before that. When war broke out in Europe in 1939, the US military was undermanned and poorly armed. Sensing that the country would likely be drawn into the growing conflict, US officials started preparing. As special assistant in the army office that oversaw construction, Groves was the point man for a massive national construction program. He managed the building of army camps and facilities around the country, including the many factories needed to produce tanks, ammunition, TNT, rifles, small arms, explosives, and other war munitions. He also learned how to manipulate the priorities and allocation system to his advantage so that he could secure the materials he needed. By the summer of 1942, with work on the Pentagon winding down, Groves had done almost everything an ambitious Corps of Engineers officer could do.

GROVES WAS IN A GOOD MOOD the morning of September 17, 1942, a typically warm and humid late-summer day in Washington, DC. A few days earlier he had been offered an attractive overseas position. It was just what he wanted. He was tired of Washington and its end-

General Leslie Groves ran the Manhattan Project from his office on the fifth floor of what is today the State Department in Washington, DC. *Courtesy of the US National Archives and Records Administration.*

less paperwork and politics. As he later said, "I was hoping to get to a war theater so I could find a little peace."

That morning he testified before the House Military Affairs Committee in Room 1310 of the New House Office Building. At 10:30, as he was leaving the hearing, he encountered Lieutenant General Brehon Somervell in the corridor. Somervell said, "The secretary of war has selected you for a very important assignment."

"Where?" Groves asked.

"Washington," replied Somervell.

"I don't want to stay in Washington."

"If you do this job right, it will end the war."

Groves suddenly realized what Somervell was talking about. "Oh, that thing."

Groves already knew about the Manhattan Project. Since that summer he had been hearing about a program that involved building massive production plants and a new kind of bomb. Groves had even been the one to propose calling it the Manhattan Project after its original headquarters in New York City.

He couldn't hide his disappointment. The project was expected to involve only about $100 million—less than he had been spending per week over the past few years. He would have to deal with a group of scientists whom he knew to be headstrong and vain. He had heard that the project was unlikely to work, and if it didn't work he would be blamed. Who knew how many times he would be testifying before Congress if that happened?

That afternoon he argued with the general who had issued the order, Wilhelm Styer. Groves said that he wouldn't accept the assignment. Styer told him that the decision could not be changed and that Secretary of War Stimson had asked specifically for Groves. Groves said that he had been promised an overseas posting. Styer said that President Roosevelt had already approved Groves's appointment.

Arguing wasn't going to work, Groves could see. He was trapped. He would have to make the best of it. Improvising quickly, he told Styer that he should be appointed to brigadier general so that the scientists would have more respect for him. Styer said that he would be promoted within the week. Groves said that he would need a higher priority rating to get materials for the project; Styer promised him that, too. Styer told Groves that he would be free to run the project however he wanted and that the War Department would give him its full support. It wasn't much, Groves knew, but it was all he would get. By the time he left Styer's office, Groves was already thinking about what he would have to do to build atomic bombs.

Chapter 10

THE B REACTOR

AS WAS THE CASE WITH MANY OTHER PEOPLE WHO MOVED TO HANFORD, Leona Woods was appalled by the dust. It got onto her face and hands and blew into her eyes and mouth. After each windstorm, a crescent of sand extended across the floorboards from the bottom of her front door on Armistead Avenue in Richland. Woods was in general unimpressed by the desert: it struck her as gray and monotonous, occupied only by rabbits, coyotes, and magpies. But on the rare occasions when rain clouds skittered across the countryside, the scent of the sagebrush was delightful.

Woods was a child prodigy from La Grange, Illinois, who graduated from high school at age 14 and earned a degree in chemistry from the University of Chicago at 19. Initially, she had asked Nobel laureate James Franck, who had emigrated from Germany to the United States in 1934, if he would be her PhD advisor. But when Franck had been a graduate student and had made a similar request of a professor, he had been told, "You are a Jew, and you will starve to death." Franck responded to Woods the same way: "You are a woman, and you will starve to death." Woods looked at the well-fed Franck and decided that she could do better for an advisor. Instead she worked with Robert Mulliken, who had served on the National Academy of Sciences committee the previous year. She was the last of Mulliken's graduate students, the others having gone to wartime jobs, and her university laboratory was "rather lonesome and empty," she later recalled. One day Herb Anderson stuck his head in the door and introduced himself. Would she be interested in coming to work for Fermi and his team? Woods was the only woman in the famous photograph of the builders of Chicago Pile 1, which might have seemed an intimidating situation. But nothing seemed to intimidate Leona Woods.

Like the other scientists working on the project, Woods worried incessantly that Germany would beat the United States to atomic bombs. "Everyone was terrified that . . . the Germans were ahead of us," she later said. "If they had gotten it before we did, I don't know what would have happened to the world. Something different. They led the civilized world of physics in every aspect at the time that the war set in, when Hitler lowered the boom. They led, not we. A very frightening time."

In 1943 Woods married John Marshall, a young physicist who had moved with Fermi from Columbia to work at the Met Lab. After Fermi's successful demonstration of Chicago Pile 1, the reactor was taken apart and moved to a laboratory about 20 miles west of Chicago, where it was renamed Chicago Pile 2. At the new lab, Woods was helping Fermi measure the rate at which various materials absorbed neutrons from the reactor, a job she thought she might lose when she became pregnant not long after getting married. She decided to hide the pregnancy beneath her baggy overalls, already bulging with a tape measure, slide rule, and notebooks. Her fellow scientists never learned that she was pregnant. But Fermi knew and was worried that he would have to deliver the baby if it arrived early, so he asked his wife for instructions on what to do. Woods later recalled:

> When he told me he was ready, it stiffened my resolution that under no circumstance would he get the chance to practice midwifery, which, in retrospect, was no doubt a disappointment to him. Anyway, the question became academic. I went to the hospital a couple of days early with high blood pressure, and came out with a baby and was back at work in a week, not quite on a par with Italian peasants, but close to it.

When Woods and Marshall moved to Hanford in 1944, their job was to "babysit the reactors," as she later described it. She was the only woman scientist there—the construction workers built her a private bathroom in the reactor building where she worked. But she could not begin her babysitting until the reactors were up and running, and by the time she arrived construction was well behind schedule. Part of the problem had been getting enough men to build the facilities. Initially, Hanford workers had their hands full just building the construction camp. They also had hundreds of miles of railroads, water and sewer pipes, and electrical lines to lay before they could even begin the production plants.

An even bigger reason for the delay was getting the facilities designed. Over the course of 1943, the engineers at DuPont and the scientists at the Metallurgical Laboratory settled into an uneasy working relationship. The Met Lab scientists, in consultation with the DuPont engineers, were in charge of the basic design of the production plants. But the detailed designs, including blueprints, came largely from Wilmington. The blueprints then were reviewed for accuracy by the Met Lab scientists, who grumbled about having to check engineers' work. The two camps argued incessantly.

Crawford Greenewalt was in charge of coordinating the work being done in Chicago and Delaware. The Met Lab scientists, he later said,

> suffered from a general disease that brilliant people, particularly in physics, seem to have; that is, because they're brilliant in their own field, they think they know everybody else's. Wigner [the leader of the reactor design team in Chicago] would have had not the slightest hesitation in telling us how to run the DuPont Company. As a matter of fact, all of the difficulty—and there was a great deal during that design—was their certainty that they knew better than we how to go at this problem, . . . as if [engineers] were glorified plumbers that had no science of their own.

The addition of Groves and the military to the mix heightened tensions. The scientists could already feel themselves forfeiting control to the DuPont engineers, whom they accused of wanting to corner the market in nuclear energy after the war. Now they could see a worse prospect on the horizon—a military takeover of all matters nuclear. A particular flashpoint for the scientists was the practice known as compartmentalization. As Groves described it, compartmentalization was at the heart of security. "My rule was simple and not capable of misinterpretation—each man should know everything he needed to know to do his job and nothing else." If someone expressed too much interest in what someone else was doing, that person was immediately under suspicion. The only one who knew what everyone was doing—and therefore the only person in a position to make informed decisions about the project as a whole—was Groves.

The secrecy inherent to compartmentalization was anathema to the Met Lab scientists. They were used to working as they did in academia—as members of open and overlapping committees. They needed to know what other people were doing to make progress, because an idea that they needed could come from anywhere. Asking them to "stick to their knitting," as Groves once described his intentions, was asking them to shackle themselves.

On this and many other issues, Groves and Szilard, who remained at the Met Lab throughout World War II, immediately locked horns. Szilard took special pleasure in "baiting brass hats," and Groves was an easy target. After one of Groves's typically overbearing presentations to the Chicago scientists, Szilard turned to his colleagues and said, "You see what I told you? How can you work with people like that?" For his part, Groves, contemplating Szilard's heavy German accent, nebulous responsibilities, and alien enemy status, suspected he was a spy. Within weeks of meeting him, Groves drafted a letter to be signed by Secretary of War Stimson ordering that Szilard be interned for the duration of the war. Stimson refused to sign, and when Groves tried to have him fired, Szilard appealed to Compton and the other scientists in Chicago to keep his job. Groves retaliated by putting Szilard under surveillance for the duration of the war. It was a comical arrangement. Szilard would sometimes lead his tails on chaotic walks through the cities he visited. Other times, he took pity on the agents and invited them to join him for a taxi ride or cup of coffee.

In other circumstances, the three-way partnership among the Met Lab scientists, the DuPont engineers, and the Corps of Engineers would never have worked. The groups were too antagonistic, too divergent in their outlooks. But all three wanted the same thing—atomic bombs as quickly as possible. As Matthias said of the arrangement, "We were a team, determined to win."

●

FINALLY, BY OCTOBER 1943, the conceptual designs were far enough along to begin digging the foundation for Hanford's first reactor. Surveyors had laid out six reactor sites, labeled A through F, strung along the bank of the Columbia River, though in the end only three reactors were needed to produce the plutonium for the Manhattan Project. The first would be built on the B site; since then, it has been called the B Reactor. The D Reactor and F Reactor would follow in quick succession.

Once the ground had been excavated and the forms set, concrete workers began pouring a foundation 23 feet thick atop the sand, gravel, cobbles, and boulders deposited over eons by the Columbia River. African American workers specialized in the pouring of concrete. "We wore rubber boots, hard hats, slicker pants, gloves to keep the concrete from messing your clothes," said Willie Daniels, a construction worker from Oklahoma who arrived at Hanford with his brother in the summer of 1943. "I knew what I was doing when it came to spreading mud." As the walls of the building rose, workers pumped the concrete through steel pipes. Their first day at work, Daniels and his brother made $19.20 between the two of them—more than his brother brought home in a month at his former railroad job.

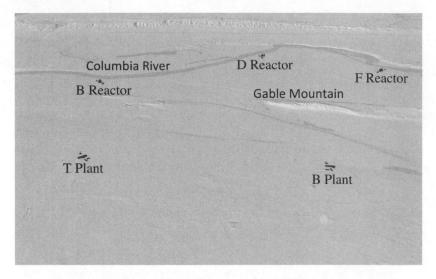

From late 1943 to early 1945, the Manhattan Project built three nuclear reactors along the right bank of the Columbia River. Two separation plants on a plateau in the center of the Hanford reservation separated plutonium from the irradiated fuel elements. On this map the outlines of structures are drawn larger than their actual sizes. The distance from the B Reactor to the T Plant is about five miles. *Map courtesy of Matt Stevenson; drawing courtesy of Sarah Olson.*

Atop the foundation was laid a steel base plate and then a layer of ten-inch-thick cast-iron blocks cooled by water pipes. These blocks were so carefully machined that their top face was level to within one two-hundredth of an inch—about the thickness of a plastic shopping bag. Along the edges of the base plate rose more cast-iron blocks, forming walls for the reactor. But these would not be nearly enough to protect people from the reactor's intense radioactivity. For that, the designers came up with an ingenious shield. It consisted of four feet of alternating steel and Masonite layers. The steel would absorb slow neutrons and gamma rays. The hydrogen atoms in the Masonite, which is a type of pressure-cooked hardboard, would slow down the fast neutrons from the reactor, which then would be captured in the steel. Together, the 4 feet of steel and Masonite were as effective as 15 feet of concrete.

By May 20, 1944, enough of the walls and roof were done to begin building the reactor itself. About a dozen construction workers at any one time were dressed in white smocks and booties, all carefully laundered in boron-free soap to keep the neutron-swallowing element out of the reactor. They began to lay down more than 75,000 graphite blocks—direct descendants of the graphite blocks Fermi and Szilard had used first in New York and then in Chicago. About four inches square and 48 inches long, each of the high-density blocks weighed 50 to 60 pounds—milling them was "about as tough as milling iron," said one of the machinists. Workers had to keep the blocks meticulously clean. Oil from the planers or drills, or even the workers' own sweat, would absorb neutrons and slow down the chain reaction. After each layer of blocks was laid, workers carefully vacuumed the graphite to make sure no contaminants remained. Layer by layer, a mass of graphite the size of a two-story house took shape inside the B Reactor.

Down the long axis of about one-fifth of the graphite blocks, the machinists had drilled a hole two inches across. As the blocks were stacked in the reactor, these holes lined up like gun barrels extending from the front of the reactor to the back. Into each of these holes, the reactor's builders inserted a long aluminum tube. Like the graphite blocks, the aluminum tubes had to be kept clean during the construction and operation of the reactor. "They had carloads of Kotex coming in—I mean a lot of Kotex, which made a lot of people wonder why," recalled one construction worker. But sanitary pads were the perfect size to keep the process tubes clean. "That's the way we swabbed all those pipes."

The design developed in Chicago and Wilmington called for workers to insert fuel elements containing uranium into each of these tubes. The uranium-235 atoms in the uranium would start fissioning, just as they had in Fermi's Chicago reactor, and the neutrons from the fissions would both keep the chain reaction going and convert uranium-238 atoms to uranium-239. After a few weeks of operation, about one of every 10,000 uranium-238 atoms would absorb a neutron. Operators would then shove fresh uranium fuel elements into the process tubes, which would cause the cooked fuel to drop out of the back of the reactor into a pool of water. After a couple of weeks to allow the uranium-239 to decay into neptunium-239 and then plutonium-239, the fuel elements would go to the separation plants to have plutonium-239 extracted from the uranium.

The uranium fuel elements would generate immense amounts of heat inside the reactor. To keep the fuel from melting, the fuel elements would take up most but not all the space within the aluminum tubes. The rest of the space would be filled with Columbia River water pumped from the front of the reactor to the back. The hydrogen atoms in the water would absorb neutrons, but it would be moving through such a confined space that each aluminum tube would hold only about a gallon of water at any given moment. The designers calculated that water would enter the reactor at about 50 degrees Fahrenheit (the Columbia is a cold river for swimming). About two seconds and 40 feet later, it would exit the reactor too hot to touch.

In addition to the holes for the process tubes, the graphite blocks had holes drilled crosswise to accommodate two sets of control rods, like the ones Fermi used in Chicago to control his first reactor. Nine water-cooled rods extended into the left-hand side of the reactor, each of which was covered with boron to absorb neutrons. While the reactor was being loaded with uranium, the inserted control rods would keep a chain reaction from occurring. Operators would then withdraw the control rods to fire up the reactor. In addition, 27 safety control rods hung above the reactor like swords ready to plunge into a stone. If the reactor seemed to be getting out of control, these rods would drop into vertical holes drilled in the graphite blocks. The reactor, when it was completed, was like a giant Jenga puzzle riddled by holes in all three directions.

A crucial design decision involved the number of process tubes in the reactor. The design from the Chicago scientists called for about 1,500 process tubes arranged in a roughly circular pattern. According to their calculations, that should be plenty to operate the reactor. But the DuPont engineers, and at least some of the Met Lab scientists,

Water towers on either side of the B Reactor provided emergency cooling water. Adjacent buildings provided water from the Columbia River and otherwise supported reactor operations. *Courtesy of the US Department of Energy.*

were worried about this number. Why not build in a safety factor by adding more tubes, they argued. Other scientists at the Met lab objected. Drilling the extra holes and fitting them with process tubes was a typical example of overconservative engineering, they said. It would slow down the project and delay the delivery of a weapon. Besides, the extra tubes weren't needed: calculations done at the Met Lab clearly showed that 1,500 tubes were enough. But on this, as on other issues, the engineers in Wilmington had the final say. By the time the blueprints made their way to Hanford, the design contained 2,004 holes for process tubes arrayed in a rectangular pattern.

Throughout the entire process of designing and building the reactor, security was extremely tight. Construction workers were cleared for one side of the reactor but not the other. They had to produce documentation halfway up a stairway, and only if their names were on a list could they pass. Security was strict for everyone. Enrico Fermi visited Hanford several times during the construction and operation of the reactors. By this time his name was Henry Farmer—all the top scientists in the Manhattan Project had aliases to disguise their identities. Once, when Fermi and Wigner, who had the alias Wagner, were coming through the gate at Hanford, a guard heard Fermi's accent and asked him if Farmer was his real name. Wigner replied, in his heavy Hungarian accent, "Farmer is his real name just as much as Wagner is my real name."

Whenever Fermi came to Hanford, he and Woods tried to find the time to go for a hike and talk. Once, sack lunches in hand, they passed through Hanford's fence and began exploring the desert. Eventually they settled down beneath an old growth sagebrush to have lunch, discussing how long ago the Columbia River had deposited the sand and gravel on which they sat. They noticed a small plane circling overhead. Soon they saw a man in the distance, holding a revolver but staring at the ground and walking a seemingly random path. He came closer until, as he rounded a nearby sagebrush, he pointed the gun at them and told them to put their hands up. Fermi was holding a sandwich in his hand while Woods held an apple in hers. "With sandwich and apple held high, not daring to take another bite," she recalled, they were marched at gunpoint back inside the fence.

EVEN AS THE B REACTOR was nearing completion, a crisis was looming elsewhere at Hanford. The uranium "slugs," as the workers called the fuel elements, could not be put into the reactor as bare metal. If they were, the water rushing down the process tubes to cool the fissioning fuel would quickly erode the uranium and carry highly radioactive fission products into the Columbia River. The slugs first had to be coated by some kind of metal. But this process of "canning" the fuel elements had become a disaster. DuPont contracted with metallurgists all over the country to come up with ways to do it, but no one could figure out how, and without fuel to load into the reactors the project would fail.

By this time, DuPont had another source of expertise on which it could draw. To try to anticipate some of the problems it would encounter at Hanford, DuPont had constructed a test reactor at the Clinton Engineer Works in Tennessee, where the Manhattan Project was building the massive plants needed to separate uranium-235 from natural uranium. Larger than Fermi's Chicago reactor but much smaller than the Hanford reactors, the X-10 Reactor rose on a small hill in the Bethel Valley about 10 miles southwest of the town of Oak Ridge. The X-10 was cooled much more simply than the Hanford reactors—with air pumped through the reactor's graphite blocks—which made it less useful as a test reactor for Hanford. But the X-10, which went critical on November 4, 1943, provided some early experience with canning fuel elements.

Still, when Fermi and Woods walked through the slug production line at Hanford on August 1, 1944—with the scheduled startup of the B Reactor less than two months away—they were shocked at how few fuel elements were ready. The problem was getting the metal casing to adhere to the uranium so that it was watertight. The casings tended to leak, rendering the fuel elements worthless. Using the most advanced technologies available, workers in the fuel fabrication plant were producing thousands of fuel elements—and almost all of them were failing the tests they had to pass to be loaded into the reactor.

This was a problem impervious to all the expertise of the scientists and engineers working on the Manhattan Project. They might be able to calculate how metals should behave in an ideal world, but they had little to no experience with how they actually do behave. The only way to solve the problem was through trial and error. Improvements, even at the last moment, were incremental. Two men figured out how to can and cap the fuel elements in a large pot of molten aluminum—a process that became known as underwater canning. The temperature of the metal baths into which the slugs were dipped had to be set unexpectedly low to get the metal to adhere to the uranium. Workers came up with new methods to test the finished slugs, with the rejects getting melted down and sent back into production.

The fuel fabrication problem was one of the most visible crises at Hanford, but many such problems, large and small, had to be solved to get the reactors to work. Scientists and engineers provided many solutions, but many came from the machinists, pipefitters, millwrights, carpenters, plumbers, and other craftworkers who built and ran Hanford. Scientists and engineers get much of the credit for innovations in industry and society, but many of those innovations actually have humbler origins. Hanford's craftworkers were deservedly proud of their contributions to its success. "We worked together to solve problems," said Cecil Gosney, who capped uranium slugs before they were loaded into the reactors. "Supervision encouraged us by listening to our ideas and suggestions. When we went home we felt we had accomplished something."

○

THE B REACTOR'S CONTROL ROOM was crowded with DuPont executives, high-ranking engineers, Met Lab scientists, and reactor operators—with some already smelling of "a drink or two of good whiskey," Woods recalled. It was the evening of September 26, 1944, the day the B Reactor would officially begin its work. From breaking ground to startup, the world's first large-scale nuclear reactor was built in less than a year.

Fermi had been experimenting with the reactor for the previous two weeks, loading increasing numbers of process tubes with fuel elements and watching how the reactor behaved. Everything had gone as planned. Without water running through the tubes, the fuel elements began to produce a chain reaction when about 400 tubes were loaded. Turning on the water dampened the reaction, but loading more tubes with fuel elements revived it. Now it was time to increase the power to production levels. About 900 process tubes were loaded with uranium. The *whoosh* of Columbia River water rushing through the reactor filled the control room.

Calculating on his slide rule, Fermi ordered the control rods withdrawn one by one. The power level began to rise. The temperature gauges in the control room—one for each of the reactor's process tubes—showed a steady increase. Soon the reactor was generating 9 megawatts of power—way below its 250-megawatt design capacity, but enough to begin producing plutonium.

As the top brass began to drift away, congratulating themselves on their tremendous achievement, an uneasy murmur began to spread through the control room. The reactor was losing power. The operators began withdrawing the control rods to keep the power steady. But all through the night the power loss continued.

By the next day, the control room was in an uproar. No matter how much the operators withdrew the control rods, the reactor continued to lose power. By 6:30 that evening, the reactor was dead. Defeated, the operators reinserted the control rods and powered down their equipment.

It was an unimaginable disaster. Hundreds of millions of dollars had already been spent on Hanford. Had the physicists and engineers overlooked something, some quirk of the physical universe that made a large-scale nuclear reaction impossible? Groves had once told Matthias that if the reactor blew up, he should "jump right in the middle of it—it'll save you a lot of trouble." This wasn't a meltdown, but it was almost as bad.

Fermi, Woods, Greenewalt, and the other scientists and engineers had immediately begun to speculate. Maybe one of the process tubes was leaking water into the graphite and poisoning the reaction. Maybe contaminants in the water were depositing on the process tubes and absorbing too many neutrons. That evening, Fermi and Woods drove back to Richland beneath a moonless sky arguing over what might have happened.

By the time they arrived back at the reactor the next morning, several of the reactor designers had come up with a plausible explanation. Overnight, one of the reactor operators had decided to pull the control rods back out to see what would happen—and the reactor had come back to life. But then it repeated its earlier performance and gradually shut itself down. Something was temporarily poisoning the chain reaction, but then that poison was wearing off. Some of the physicists had wondered about this possibility, but they had not been able to test it. When uranium fissions, it produces a wide variety of fission product elements, which then radioactively decay into yet more elements. Maybe one of these radioactive decay products was capturing neutrons and squelching the chain reaction. The physicists checked their tables of radioisotopes, and one immediately caught their eye. Xenon-135, which is one of more than 30 radioactive isotopes of xenon, has a half-life of about nine hours, which was suspicious given how the reactor was behaving. Physicists did not then know xenon-135's capacity to capture neutrons, but it later turned out to have the highest neutron absorption rate of any isotope. Testing over the next few days confirmed their suspicions— xenon-135 being produced in the fuel elements by the fissioning of uranium was shutting down the reactor.

Could anything be done about it, or was the reactor ruined? This is where DuPont's decision to add the extra process tubes turned out to be an inspired hedge. In the days after the crisis, the scientists and engineers at Hanford calculated that if they filled all the process tubes with uranium fuel, the extra neutrons would overcome the xenon-135 poisoning effect. As originally designed, with 1,500 tubes, the reactor would not have been large enough to do the job. But the extra 500 tubes gave it the capacity it needed.

DuPont's insistence on playing it safe in the reactor design led to a bit of doggerel that was repeated at the company for decades. "Marse George" in the sonnet is George Graves, one of the DuPont engineers who pushed for the extra process tubes:

> *The tale's been told, as well you know*
> *That Hanford nearly flopped, although*
> *The piles were later made to go*
> *Through brilliant engineering.*

The B Reactor, which has changed very little since World War II, is part of the Manhattan Project National Historical Park. © *Harley R. Cowan—All Rights Reserved.*

The reason they were made to run
Was that a battle had been won
Long months before in Wilmington
With brains and persevering.
We'd cobbled up a tight design
Hewed strictly to the longhair's line
To us, it looked mighty fine—
A honey we'd insist.
But Old Marse George, with baleful glare
And with a roar that shook the air
Cried "dammit, give it stuff to spare
The longhairs might have missed!"
And later when the crisis came
Twas George's trick that saved the game.

The reactor would work, but not as smoothly as anticipated. Loading all 2,004 process tubes permitted the reactor to run at full power. But the xenon poisoning made the reactor a persnickety machine. If the reactor shut down for some reason, operators had to get it back online in a half hour, or the buildup of xenon would force them to shut it down for half a day.

Controlling the reactor was also tricky because of the xenon building up and then decaying in the fuel elements. But the operators were resourceful, and by December they had figured out how to position the control rods to keep the power levels high. Meanwhile, the virtually identical D Reactor, a few miles farther down the Columbia, began operating in December, and the F Reactor came online in February. By spring 1945, all three reactors at Hanford were making plutonium.

But converting uranium to plutonium inside a fuel element was just the first step. Before being placed in an atomic bomb, the plutonium had to be extracted from what were, after spending a few weeks inside a nuclear reactor, among the most dangerous objects on Earth.

Chapter 11

THE T PLANT

LEONA WOODS WAS ENTRANCED. FROM THE FUEL ELEMENTS SCATTERED across the bottom of the pool rose a ghostly blue light. It was roughly the blue of the sky but more vivid, electric—the light shimmered in the gently moving water like a living thing. The Soviet scientist Pavel Cherenkov had predicted about a decade earlier that charged particles moving through a medium like water would generate a shock wave of visible light. But this was the first time that Cherenkov radiation had been clearly seen.

After spending several weeks being bombarded by neutrons, the fuel elements in the pool behind the B Reactor's core contained what Greenewalt called a "dog's breakfast" of radioactive fission products. Some of these fission products decay quickly and are no longer dangerous after a few days. But the ones with longer half-lifes, like strontium-90 and cesium-137, remain dangerous for decades or centuries.

At Hanford the fuel elements normally stayed at the bottom of the reactor pool for several weeks while uranium-239 decayed to plutonium and the radiation from the short-lived fission products abated. Operators then used long tongs to reach to the bottom of the pool and place the irradiated fuel element into buckets. They maneuvered the buckets into lead-lined casks, raised the casks from the bottom of the pool, and placed them on flatbed rail cars. From there the fuel elements traveled five miles south to what were certainly the strangest facilities ever built at Hanford. On Hanford's central plateau, just

south of a sharp fin of basalt known as Gable Mountain, construction workers had erected three immense concrete monoliths. Known
as the T Plant, B Plant, and U Plant, in the order in which they were
completed, the workers called them "Queen Marys" because of their
size, though they looked nothing like ships. The separation plants
were featureless gray cuboids almost 900 feet long, 65 feet wide,
and 85 feet tall. They had no windows, few doors, and no adorn-
ment. From the inside they looked like long concrete canyons, and
the name stuck.

Essentially, these buildings would reproduce, on an immense
scale, the tabletop chemical processes that Glenn Seaborg and Art
Wahl had used to discover plutonium three years before. When Seaborg took the train from Chicago to Hanford at the end of May 1944,
he wrote in his diary that it was "an awe-inspiring experience to see
the thousands of workmen busily engaged in the building of these
complicated edifices." Down the length of each canyon building ran
about 40 cells with thick concrete walls, like a long, single-row egg
carton. Each cell, topped by a 35-ton concrete lid, contained a different piece of equipment, depending on which step of the process the
cell handled. "One sees nothing but 860 feet of valves, meters, indi-
cators, controls—a fantastic sight," Seaborg marveled.

On December 26, 1944, the first irradiated fuel elements arrived
at the T Plant. There the operator of an overhead crane unloaded
the slugs, which the workers called lags after they were irradiated,
and placed them into a massive steel tank. Another operator then
filled the tank with sodium hydroxide, which dissolved the cladding around the uranium metal. Once the dissolved cladding was
drained away, the tank was refilled with nitric acid, which dissolved
the uranium. The result was about 3,000 gallons of liquid feed—
enough to fill a small backyard swimming pool. Dissolved in this
feed material was less than a tablespoon of plutonium—literally
one part in a million. The task now was to separate this plutonium
from everything else.

The feed material began traveling down the canyon building. In some cells, operators added chemicals that caused the uranium and fission products to become solids while the plutonium remained a liquid. In centrifuges as large as a man that spun 15 to 30 times a second, the uranium and fission products moved to the outside of the bowl while the liquid remained in the center. This plutonium-bearing liquid then traveled through a pipe to the next cell. There operators added chemicals that turned the plutonium to a solid, which was separated out in another centrifuge. A solid, a liquid, spinning, scraping—as the dissolved fuel elements moved down the canyon buildings, the plutonium became more and more concentrated.

All of this had to be done by remote control. Once the initial batch of fuel elements went through a canyon building, the processing cells became too radioactive to enter. Behind massive concrete walls, the crane operators looked through periscopes to manipulate their equipment. A television camera mounted on the bridge crane sent a picture to the control room—it was the first use of closed-circuit television in history. Crane operators quickly became experts at lifting the five-foot-thick concrete lids off cells and changing pipe fittings with remote-controlled hooks and impact wrenches. The canyon buildings had no pumps to move fluids from one cell to the next. "We wanted as few moving parts as possible," said the head of the design team at DuPont, Raymond Genereaux. "Moving parts have problems. They jam, wear out, fail. What we used were steam jet ejectors." Centrifuges were designed to make noise if they were malfunctioning. "That helped," Genereaux said. "If you heard a rough noise, something was wrong. If it was humming you were okay."

The end product of this process, which came to be known as reprocessing (because the uranium had already been processed in the reactor), was a dribble of almost completely pure plutonium. Plutonium-239 is not very radioactive—you can hold a chunk of it in your hand. But the alpha particles it gives off generate heat as they slow down, which gives plutonium a disquieting warmth. This was the substance that chemists, in a finishing shop next to the T Plant, prepared to ship to the bomb makers.

Atomic Bomb Plant — H.E.W. Process Bldg.
Photo by Robley L. Johnson

The separation plants at Hanford, also known as canyon buildings, removed plutonium from irradiated uranium using the chemical processes developed by Glenn Seaborg and his colleagues. *Courtesy of the US Department of Energy.*

THE CREATION OF PLUTONIUM at Hanford generated fantastic amounts of gaseous, liquid, and solid waste. Woods recalled that whenever a new batch of fuel elements arrived at the separation plants for reprocessing, "great plumes of brown fumes blossomed above the concrete canyons, climbed thousands of feet into the air, and drifted sideways as they cooled, blown by winds aloft." The plumes consisted in part of smog from the nitric acid used to dissolve the uranium, but more ominous substances also went up the stacks. Some fission products are gases, like the xenon that poisoned the reactor. Other elements are so volatile that they got swept up in the exhaust air coursing through the canyon buildings, including radioactive isotopes of iodine, ruthenium, strontium, and cerium. Hanford's operators tried to release radioactive gases only when the wind was blowing strongly enough to disperse them widely. But Hanford sits in a low basin. When the winds were calm, the air and its toxic contaminants sat in the basin like soup in a bowl. When the wind was blowing, plumes of radioactive gases crested the surrounding hills and headed toward downwind fields and towns.

Another form of waste was water that had passed through the reactors. In the two seconds it took for Columbia River water to travel through the reactors' process tubes, the intense neutron flux could make almost any contaminants radioactive, including calcium, chromium, and zinc. If fuel elements in the process tubes leaked, uranium and its fission products also entered the water. The water from the reactors cooled off in holding basins for a few hours, allowing its short-lived radioactivity to decay. But then it flowed back into the Columbia River, carrying the longer-lived radioisotopes with it.

Many of the other facilities at Hanford also generated radioactive wastewater, including the fuel fabrication facilities and separation plants. At first, operators directed this water into shallow depressions in the landscape, which turned into muddy swamps. But measurements showed that these swamps were getting increasingly radioactive. When the swamps dried in the spring and summer, the wind picked up radioactive dust and blew it across the landscape. To prevent this problem, construction workers began building reverse wells, French drains, and timber-lined cribs into which water could flow. In all these cases, the idea was to get the water underground where its radioactive contents would do less harm. But water sent underground, along with its contaminants, rarely stays in one place.

Operations at Hanford also generated huge amounts of solid waste: boots, gloves, worn-out reactor parts, tools, cardboard, soil, glassware, wire, plastics, the carcasses of radioactive animals—enough waste, eventually, to completely fill more than 4,000 railroad boxcars. (In fact, the waste included boxcars.) This waste went into hundreds of landfills, trenches, tunnels, and dumps scattered around the site. Hanford is a big place, and wastes that would quickly overflow a more crowded industrial site could be discreetly tucked away. But when rain fell, it percolated through the soil covering the waste and then through the waste itself, carrying radioactive particles deeper into the earth.

One last form of waste was the worst. The separation plants used immense quantities of chemicals to dissolve the fuel elements and separate them from other elements. Once these chemicals were used in a canyon building, they generally were too radioactive to use again. To dispose of them, the radioactive chemicals were channeled through pipes to gigantic underground tanks built near the separation plants. Eventually, 177 tanks were built at Hanford to hold the high-level radioactive waste from manufacturing plutonium—each as big as an auditorium. The initial tanks had lifetimes estimated at 20 years, after which the builders of Hanford figured that someone would have come up with a means of permanently disposing of the intensely radioactive chemicals. More than three-quarters of a century later, the wastes continue to sit in their tanks beneath the desert sands.

◦

GROVES, MATTHIAS, GREENEWALT, and other army and DuPont officials knew that the workers at Hanford would have to be protected from the radiation it produced. They therefore set up departments responsible for health and safety, just as they had in Chicago. As in Chicago, DuPont established standards for radiation protection and required that workers wear radiation monitors. As in Chicago, these rules provided sufficient protection most of the time but not always.

One dilemma at wartime Hanford was that most employees, for reasons of security, were not supposed to know that they were working around radiation. Managers therefore urged employees to follow safety practices without explaining exactly why. "Hesitate, Cogitate, Be Safe," billboards proclaimed. In a cartoon distributed to workers, Safety Sam exhorted his fellow workers to "report or eliminate hazardous conditions promptly." Safety inspectors and health personnel known as radiation monitors kept an eye on workers and corrected unsafe practices. An elaborate system of code words took shape, where uranium was "base metal," plutonium was "product," and radiation was "activity."

Researchers at Hanford continued the work begun at Chicago on the health effects of plutonium and other radioactive substances. By this time, health workers were becoming increasingly concerned about the harmful effects plutonium could have on biological tissues. It was not a problem outside the body, but if small particles were inhaled or ingested, or entered the body through cuts, they could lodge in bones, liver, or lungs and emit radiation throughout a person's life. This radiation is not necessarily harmful. We all have many radioactive particles in our bodies, and external sources of radiation continually damage our cells. But experiments in animals showed that very small amounts of plutonium could cause cancer, requiring workforce standards that had to be rigorously enforced.

Workers generally followed safety procedures, but not always. "We did a lot of foolish things," said one chemist who worked at Hanford during the war. "We took risks to get results rather than spend days and weeks designing equipment so that you could do everything without getting very close or taking any risks. . . . We felt a tremendous pressure. We believed the Germans likewise were doing this. And we had to beat the Germans because if they got there first they would win the war."

EVEN AS THE T PLANT began to produce plutonium at the beginning of 1945, another major construction project was nearing completion. Groves and Matthias had always thought of the Hanford camp as temporary. It would accommodate the rowdy construction workers for a few years and then be abandoned. But the people who would be overseeing and operating the production plants—the construction managers, scientists and engineers, DuPont executives, plant operators, security officers—needed a place to live. The nearby towns of Kennewick and Pasco were too far away, too provincial, and too accessible to house the high-level employees of a secret facility. Groves needed a town that would attract people to the middle of a desert, keep them there, and isolate them so that no one else knew what they were doing.

At four o'clock in the morning on March 2, 1943, the renowned Spokane architect Gustav Pehrson received a phone call. A committee had recommended that Pehrson bid on a big project going up about 150 miles southwest of Spokane. "Impossible," he said, and hung up. When the phone rang again, it was Matthias on the line and his tone was different: "You *will* meet Lieutenant Colonel Kadlec at 2:00 p.m. tomorrow in Pasco." Pehrson inspected the site the next day, and two weeks later his bid was accepted. He and his firm were to provide plans for a completely new town to be built atop the small farming community of Richland—streets, houses, dormitories, parks, commercial buildings, and infrastructure. The purpose of the town was shrouded in mystery, but it would have to house thousands of people. Pehrson's initial designs were due in a week.

Over the course of several conversations with DuPont officials, Pehrson developed his ideas, despite not knowing exactly what the residents of his new town would be doing. The designer of some of Spokane's most distinguished downtown buildings, the Swedish-born Pehrson envisioned a town of comfortable houses on large lots set back from gracefully curving streets. He designated his two dozen or so house designs by letters of the alphabet. Even today, people in Richland will say that they live in a B house—an 882-square-foot single-story duplex with two bedrooms and a half basement—or an L house—a 1,536-square-foot two-story structure with four bedrooms and bathrooms upstairs and downstairs. Pehrson designed the houses with large plate glass windows so occupants could look out at the neighborhood and the endless desert sky. He gave them modern appliances and good ventilation to stay cool in the scorching summers. The houses would be surrounded by trees and shrubs for privacy and shade. Residential streets fed into a commercial core of shops and public facilities, as in traditional American small towns. Originally, the types of houses were supposed to be entirely mixed so that all neighborhoods had people at different income levels. But as the plans evolved, sites for the larger and nicer houses tended to move toward the Columbia River while the smaller houses and duplexes clustered in the dusty plains beyond.

When Groves got his first look at the designs, he exploded. These were supposed to be residences for wartime employees, not comfortable middle-class homes. To Groves, it looked as if DuPont was using federal funds to build a luxury resort in the middle of the desert. How would that look when Groves appeared before congressional committees after the war to explain how he'd spent the public's money? He immediately started making cutbacks, so that the houses in Richland looked simpler, more like the houses he had grown up in. He took out the picture windows and elaborate molding. He scratched the men's clothing store and funeral parlor from the commercial center. He even renamed the planned hotel the "transient quarters," since "hotel" sounded too extravagant to him.

DuPont and Pehrson fought back. They pointed to "the necessity for maintaining high morale among workers transplanted to what will probably seem a strange country," as Pehrson put it in his November 1943 plans, which "cannot be achieved by crowding skilled and veteran workers into inadequate dwellings." Grudgingly, Groves gave way, at least on a few points. Houses could have three or four bedrooms, not the one or two Groves preferred. The commercial center could contain more businesses than he wanted.

Over the next 18 months, DuPont's contractors built more than 4,000 houses using Pehrson's designs. The result was an odd combination of government town, company town, and detention center. The government provided everything—furniture, light bulbs, electricity, a lawn mower. Yet residents quickly adopted the language and imagery of the West, emphasizing their own volition and independence. "Hanford workers were living at the frontier," historians John Findlay and Bruce Hevly have observed. "They and their families had left the smoky industry of the East behind and moved West into a nuclear-powered new day."

Richland was something new on the American landscape. The

majority of the people living there were blue-collar workers. They manufactured fuel elements, adjusted controls on the reactors, or tested the intermediate products of the separation plants. Yet they were able to fashion new lives for themselves, far from the ones they had left behind. Five years before Abraham Levitt and his two sons, using construction techniques adopted from the navy, built Levittown on Long Island, the nation's first mass-produced suburb rose on a dusty, windswept plain in south-central Washington State.

Richland gave its residents incomes and access to amenities that they never could have achieved before the war, but they paid for their privileges. They were under continual surveillance and were severely limited in what they could say or do. Employees underwent background investigations, fingerprinting, and regular vehicle searches. Security officers known as Groves's creeps listened to conversations, tapped phone calls, edited outgoing mail, and conducted investigations. To catch anyone who might be giving away secrets, agents posed as hotel clerks, tourists, electricians, painters, and even gamblers. Forty-four documented cases of "un-Americanisms" involved such infractions as possessing a two-way radio or criticizing President Roosevelt. If a spouse or child created a disturbance, an employee could lose a security clearance, and spreading rumors warranted an immediate reprimand or termination. Today, people are used to government and large corporations having access to large amounts of personal information. In Richland and the other cities built and run by the Manhattan Project during World War II, this surveillance society was something new.

Chapter 12

IMPLOSION

ON FEBRUARY 2, 1945, FRANKLIN MATTHIAS GOT INTO A CAR IN Richland carrying a square wooden box. He drove past the Horse Heaven Hills, through the spectacular windswept gorge that carries the Columbia River through the Cascade Mountains, and into the city of Portland. There he boarded the West Coast passenger train for Los Angeles, the box and a suitcase at his side.

At Los Angeles Union Station, he met a security officer from the Manhattan Project's Los Alamos laboratory. "Do you have a locked compartment to go back to Los Alamos?" he asked, nodding at the box.

"No, I couldn't get a bedroom," answered the officer. "I have an upper berth."

"Do you have any idea what we have here?" Matthias barked. "You better highball it down and get yourself a locked compartment."

"What do you mean?"

"I mean that it cost $350 million to make it. If you don't get it to Los Alamos intact, you're going to be in a hell of a lot of trouble."

THE NEXT MORNING, the first shipment of Hanford plutonium arrived at Los Alamos. By this time, all three of the Manhattan Project's major sites were running around the clock. At Hanford, the reactors and separation plants were churning out product. At Oak Ridge, gigantic factories nestled amidst the Tennessee hills were separating uranium-235 from uranium ore. And at Los Alamos, scientists and engineers under the direction of Robert Oppenheimer—Seaborg's colleague at Berkeley, whom Groves had chosen to run the lab— were designing weapons that would produce the most powerful explosions ever created by humans. But Los Alamos was also in the midst of the Manhattan Project's greatest crisis, and the reason for that crisis was plutonium.

The initial shipment of plutonium from Hanford in February 1945 was not the first batch of plutonium to arrive at the laboratory. The previous year, Seaborg's Berkeley colleague Emilio Segrè, whom Oppenheimer had recruited to Los Alamos, had received a couriered shipment of plutonium from the pilot reactor in Tennessee. Immediately, he tested it to see whether it was going to cause a problem that he and Seaborg had been worrying about for months. The results of the test were disastrous. The entire approach to making an atomic bomb with plutonium would have to change.

Physicists had come up with a relatively simple design for an atomic bomb not long after the discovery of fission. It hinged on a concept known as critical mass. A free neutron moving inside a mass of either uranium-235 or plutonium-239 can do one of two things. It can strike the nucleus of another atom, in which case that atom is likely to fission. Or it can escape from the sides of the mass and fly away without causing another fission.

In a small mass of plutonium, a neutron is more likely to escape than it is to hit another atom, because the sides of the mass are always nearby. But that changes as a piece of plutonium gets larger. At a certain mass, known as the critical mass, a neutron is more likely to hit another plutonium atom than to leak into space. At that point, the chain reaction begins to grow exponentially, releasing immense quantities of energy.

But even with a critical mass of plutonium, not all the plutonium atoms will fission. A plutonium atom takes about a billionth of a second to split and release neutrons, and in a critical mass of plutonium these neutrons take a few more billionths of a second to find and split other plutonium atoms. As a result, the chain reaction occurs very quickly—an atomic bomb releases all the energy it's going to release in less than a millionth of a second. But before all the atoms in a critical mass can split, the energy released by fission blows the bomb apart. When that happens, the neutrons from a fissioning atom can no longer find other plutonium atoms to split. In the plutonium bombs designed at Los Alamos, only about one-seventh of the plutonium atoms split before the bomb blew itself apart and the chain reaction ceased.

The concept of critical mass suggested a basic bomb design. A cannon barrel would have two pieces of uranium-235 or plutonium-239, both smaller than a critical mass, at either end. One piece would be shaped like a stack of washers, forming a cylinder with a hole in it. The other piece would be a cylindrical post just small enough to fit inside the hole. A conventional explosive would send the hollow cylinder hurtling down the cannon's barrel. As the hollow cylinder settled around the solid cylinder, like a ring slipping onto a stubby finger, the combination of the two pieces would form a critical mass and explode.

But the designers were always aware of a potential complication. When the hollow cylinder began to slip over the solid cylinder, the two pieces would form a critical mass before they were fully joined. A spare neutron anywhere in this assembling mass would almost instantaneously set off a chain reaction. As a result, the bomb would blow itself apart well before the two masses had completely merged. The result would be what the designers called a fizzle—a premature detonation far less powerful than the explosion they were seeking.

The solution to this problem was to make sure that the two masses did not contain any free neutrons until they were fully joined. That required paying attention to something called spontaneous fission. Uranium-235 and plutonium-239 are so close to the edge of stability that they sometimes fission all by themselves, without being hit by a neutron. When they do, they release their usual two or three neutrons. If they were to do this within a critical mass, the neutrons would start a chain reaction. That would be a calamity if it happened in a gun-type bomb when the two pieces were partly but not fully joined. The result would be a relatively small explosion that would destroy the bomb without having much of an effect.

The bomb designers were confident that spontaneous fission would not be a problem for uranium-235. Its rate of spontaneous fission is so low that a fission would be very unlikely to occur during the fraction of a second while the two pieces of material were joining. The same was true of plutonium-239. Its rate of spontaneous fission is greater than that of uranium-235 but not enough to cause a fizzle.

Seaborg and Segrè were worried about something else. What if plutonium-239, after being created in a reactor, were to absorb another neutron and become plutonium-240? The former Berkeley colleagues had reason to believe that plutonium-240 would have a much higher spontaneous fission rate than plutonium-239. Furthermore, there would be no practical way to remove the plutonium-240 from the plutonium-239. If nuclear reactors inevitably ended up producing significant quantities of the heavier isotope, the entire plutonium production project could be a bust.

That's what Segrè was testing when he got his first batch of plutonium from Oak Ridge. He and Seaborg had been right to worry. The plutonium contained way too much plutonium-240 for the gun-type design to work. And if the gun design wouldn't work, the hundreds of millions of dollars being spent at Hanford would be wasted.

But not all was lost. By then the bomb designers in Los Alamos knew about another way to create a critical mass of plutonium, a better way. What if explosives could be set off around a subcritical sphere of plutonium so that it was suddenly compressed? Because of the sphere's increased density, neutrons from a fissioning atom would need to travel less distance to find another atom to fission. As a result, neutrons would be less likely to escape from the sphere before fissioning another atom. The compression would happen so quickly that neutrons from plutonium-240 probably would not have time to generate a fizzle before the full nuclear explosion occurred.

The idea, known as implosion, might be a way to salvage the use of plutonium in atomic bombs.

A small research group at Los Alamos had been working on the concept even before the plutonium-240 crisis. But implosion seemed so technically difficult that it remained on the fringes of the laboratory's efforts. The problem is that explosions don't work that way. If you throw a pebble in a pond, it generates waves that travel outward in a circular pattern. Implosion requires the opposite: a spherical wave traveling inward. How can you turn an explosion outside-in?

In 1944 the spontaneous fission problem gave the bomb designers no choice. They had to develop implosion to get plutonium to work in a bomb. Oppenheimer responded by almost completely reorganizing the Los Alamos laboratory. By that fall, hundreds of scientists, engineers, and technicians were working on implosion. The canyons around Los Alamos resounded with exploding charges as researchers tried to figure out how to direct an explosion inward rather than outward.

The critical breakthrough was the idea of using shaped explosives. Using commercial candy-making machines, the bomb designers created explosives of different sizes, shapes, and compositions. They then put these explosive lenses together in spherical configurations designed to focus the force of an explosion inward. When detonators on the outside of the sphere simultaneously ignited the explosives, a shock wave traveling inward was bent by the shape of the explosives and the speed with which they burned. The goal was to create a converging wave that would compress a small sphere of plutonium at the bomb's core, like squeezing a snowball in your fist.

But getting the idea to work was fiendishly difficult. No theory satisfactorily explained the forces generated by shaped charges. Instead, the Los Alamos scientists had to experiment. They devised ingenious techniques to observe what was going on during test explosions. They also used some of the world's first electronic computers to approximate the answers to equations that they could not solve by hand. When voids formed in the test explosives while they were being cast, the researchers carefully drilled through the explosives using dental tools and filled the holes with small amounts of explosive melted in a steam-heated pot. When asked about the safety of this procedure, the leader of the implosion team, George Kistiakowsky, said, "You don't worry about it. If fifty pounds of explosive goes [off] in your lap, you have no worries."

The invention of implosion at Los Alamos required a host of other innovative technologies. Detonators on the outside of the explosives had to fire at exactly the same time to generate a smooth shock wave. A device known as an initiator, placed at the very center of the bomb, had to release neutrons into the imploding core at the precise instant when they were needed to trigger a chain reaction.

Gradually the Los Alamos researchers made progress. Images of test explosions showed that implosion was becoming more symmetric and powerful. The metallurgists at Los Alamos figured out how to alloy and cast plutonium so that implosion would squeeze the material uniformly. By the end of February 1945, Oppenheimer was optimistic enough to say, "Now we have our bomb."

The development of implosion altered the course of world history. An implosion bomb requires much less nuclear material than a gun-type bomb. The bomb dropped on Hiroshima, which was nicknamed Little Boy for the long thin cannon barrel it contained, used about 140 pounds of uranium from the isotope separation facilities in Oak Ridge, Tennessee. The bomb dropped on Nagasaki, nicknamed Fat Man because of the spherical explosives it contained, used just over 13 pounds of plutonium—less than one-tenth as much material. In fact, if spontaneous fission had not been a problem and implosion had never been developed, Hanford would not have been able to produce enough plutonium for a gun-type bomb by the summer of 1945.

The physicists in Los Alamos were confident that a gun-type bomb using uranium-235 would work, but the amount of material required posed a big problem. Even in the spring of 1945, the separation of uranium-235 from natural uranium ore was not going well. Groves would be lucky to have enough uranium for a single gun-type bomb by that summer, and he wouldn't have enough for another gun-type bomb before the end of the year. Even if the Los Alamos scientists had wanted to test the gun-type bomb, they didn't have enough uranium to do so.

The possibility of making an atomic bomb with one-tenth as much fissile material changed everything. If implosion worked, Hanford should be able to manufacture enough plutonium for a bomb by the middle of the summer. After that, Hanford should be able to produce enough plutonium for several bombs per month. If the Manhattan Project had been forced to stick with gun-type bombs, the United States might have had a single atomic weapon by the summer of 1945. With the development of implosion, it could make as many bombs as it wanted.

Still, implosion was far from a sure thing. The scientists and engineers at Los Alamos were not at all confident that they could get it to work. They certainly couldn't see setting off the very first implosion-based bomb in combat. They would need to test one first. Groves immediately objected. He needed a bomb as soon as possible. The project could not waste plutonium on an unnecessary test.

The bomb makers persisted. The United States could not risk dropping a dud, and Hanford would be able to make enough plutonium for a second bomb in a couple of weeks. Groves knew they were right about the test. If he dropped a bomb on Japan that didn't work, his reputation would be ruined. Plus, he was thinking by this time about what would happen after the war, when the United States would be the only country with nuclear weapons. In 1944, a Manhattan Project scientist had been startled to hear Groves admit that the development of atomic bombs was really meant to subdue the Soviet Union, because the United States would have nuclear weapons after the war and the Soviets would not. If implosion could be developed, it would make possible a virtually unlimited supply of atomic bombs. Yes, a test was a small price to pay for access to such a valuable technology.

○

AS THE ALLIES PREPARED for the invasion of France in the late spring of 1944, Crawford Greenewalt became increasingly worried that the Germans would use atomic bombs to fend off the Allied troops. "I don't think I'll ever forget D-Day," he recalled many years later. "I was actually afraid to look at the newspapers, because with our troops concentrated on the beaches of France, and with the enormous concentration of men and material, I thought, 'Well, if the Germans really have it, here it comes.'" Groves was worried about something else. If the Germans had built a reactor and were using it to make plutonium, they could produce large quantities of radioactive materials to drop on invading troops. He convinced Eisenhower to have several soldiers carry radiation detectors onto the Normandy beaches on June 6, though no radiation or any other nuclear devices were detected.

Groves never had any hard evidence that the Germans were working on an atomic bomb. The US government based the entire Manhattan Project on hearsay, speculation, and worst-case scenarios. Nevertheless, Groves did not have good evidence that the Germans were not working on a bomb, and as the Allied troops made their way deeper into Europe, he continued to worry about German nuclear weapons. Earlier, he had approved the establishment of a covert counterintelligence group that came to be known by the name Alsos, though Groves was infuriated when he learned about the name, since *alsos* is the Greek word for a grove of trees. The civilian component of the operation was headed by Samuel Goudsmit, an accomplished nuclear physicist from the University of Michigan. Born in the Netherlands and fluent in several languages, Goudsmit had been working on radar during World War II and knew little about the Manhattan Project. That appealed to Groves—he would have little to tell the Germans if he were captured. Goudsmit also had plenty of incentive to fight. In March 1943, he had received a letter from his Jewish parents, who had been living in The Hague before the war. Instead of a return address in the Netherlands, the envelope bore the site of a Nazi concentration camp. Its purpose: to bid their son farewell.

After D-Day, the Alsos team followed the Allied troops into Europe, looking for any hint of a German atomic bomb. Records gathered in Brussels indicated that Germany had procured large shipments of uranium in 1942 and 1943, which greatly concerned Groves. Could the Germans, even as their war-fighting ability crumbled, be working on one last stand?

Further documents indicated that Strasbourg had been one location of German nuclear research. As soon as the US Army entered the city in late November, Goudsmit followed. There, he worked for four days and nights by candlelight, under oppressive air raids, reading the captured papers of University of Strasbourg physicist Carl Friedrich von Weizsäcker. Here was just what he needed. The papers revealed that the Germans had been unable to extract large quantities of uranium-235 from their materials. They had tried to build a reactor in 1944, but they had failed to produce a chain reaction. "The conclusions were unmistakable," Goudsmit later reflected. "The evidence at hand proved definitely that Germany had no atomic bomb and was not likely to have one in any reasonable time." He and his Alsos team related the recovered information in "The Strasbourg Report," which they jubilantly shipped back to Groves.

Groves wasn't satisfied. He considered the Strasbourg information too easily gathered and wondered if it had been planted. Furthermore, the information meant that the German scientists had been working on a bomb, which meant that they still could be working on a bomb somewhere inside Germany. He ordered the Alsos investigators to keep looking.

Throughout the first part of 1945, the Alsos team followed the army as it penetrated deeper into Germany. In Stadtilm, they found eight tons of uranium oxide and documents describing the German nuclear program. They began to capture scientists who had worked on nuclear energy during the war, though none provided much information. In the small town of Haigerloch, an Alsos team found a small experimental nuclear reactor made of graphite blocks, though it did not contain any heavy water or uranium. Only on April 23, after the capture of 1,200 tons of uranium ore hidden in caves near Strassfurt, was Groves able to write to Army Chief of Staff George Marshall: "The capture of this material, which was the bulk of uranium supplies available in Europe, would seem to remove definitely any possibility of the Germans making any use of an atomic bomb in this war."

Compared with the Manhattan Project, the German nuclear program was laughably amateurish. "The whole German uranium setup was on a ludicrously small scale," Goudsmit later wrote. "All it amounted to was a little underground cave, a wing of a small textile factory, a few rooms in an old brewery. . . . Sometimes we wondered if our government had not spent more money on our intelligence mission than the Germans had spent on their whole project."

The German nuclear program never got off the ground for a variety of reasons. An early measurement that neglected to account for impurities indicated that graphite would make a poor moderator in a reactor. Instead, the Germans decided to use heavy water as a moderator, which was so technically difficult that they never did get a reactor to work. Other problems were administrative. The Germans never found someone like Groves to unify and lead the program forward, and much of the top scientific talent in Germany had fled to the United States to avoid Nazi persecution. Hitler was obsessed with other weapons, including the V-2 ballistic missile, and never embraced a nuclear program, as Roosevelt had done. And, as nuclear historian Alex Wellerstein has observed, German scientists never believed that the United States would be capable of building atomic bombs; they "lacked the fear of an Allied project that the Allies had of them."

The US military never got to the point of considering where or how atomic bombs might be used in Europe. Nor was the idea of dropping such bombs on Europe ever thought through carefully.

Would a US president be willing to destroy European cities or large portions of the countryside with a single bomb? And if a weapon were dropped and failed to detonate, the Germans would have a ready supply of uranium, manufactured in Oak Ridge, or plutonium, manufactured in Hanford, to use in their own bomb.

But even if the German nuclear program had never been a serious threat, the war was far from over in early 1945. In the Pacific, US forces were struggling to take Iwo Jima and then Okinawa. If those battles were successful and the Japanese still refused to surrender, the Allies might need to invade the Japanese mainland. Even though atomic bombs were not needed against Germany, they could prove useful elsewhere.

In early 1945, when he finally had demonstrated that the German atomic bomb project had made little progress, Goudsmit remarked to his Alsos colleague Robert Furman. "Isn't it wonderful that the Germans have no atom bomb? Now we won't have to use ours."

"Of course you understand, Sam," replied Furman, who had worked with Groves before the war, "if we have such a weapon, we are going to use it."

Chapter 13

WASHINGTON, DC

GENERAL GROVES NEEDED MORE PLUTONIUM. IN THE SPRING OF 1945,
Hanford was not producing enough material for both a test bomb
and bombs to use in the war. The operators were still learning how
best to work their massive reactors and separation plants. For Groves
to have enough nuclear material for more than one bomb before the
end of the war, Hanford had to work faster.

On March 24, he arrived in Richland with a contingent of DuPont
engineers, and "a discussion was held as to the production schedule
to be followed in the coming few months," Matthias wrote in his
journal. Groves and Matthias had three possible ways of speeding
up plutonium production. First, they could run the reactors at more
than their design power, which would convert more uranium in
each fuel element into plutonium. Second, they could run more fuel
through the reactors, increasing the amount of uranium exposed
to the reactors' neutrons. Third, they could reduce the cooling-off
period before the irradiated fuel elements entered the separation
plants, allowing the plutonium to be extracted sooner.

By the beginning of April, Groves and Matthias were using all
three approaches. "Production results have been extremely good,"
Matthias wrote of what came to be known as the speedup, "and the
time of delivery of units has been faster than originally planned."
But the speedup exacted a severe environmental toll. Uranium that
spent less time in the reactor contained less plutonium, so more of it
had to be processed. The streams of radioactive chemicals flowing

into the steel tanks near the processing plants swelled. Worse, cooling the fuel elements for just a few days gave the radioactive fission products less time to decay. The levels of radioactive gases billowing from the stacks next to the separation stacks skyrocketed.

Matthias had long since given up hand-delivering the plutonium to Los Alamos. It now traveled in olive-green panel trucks that looked like ambulances. Convoys of five vehicles and 10 men equipped with shotguns, revolvers, and machine guns traveled from Richland through Boise to Salt Lake City, where another group met the Hanford group to take the plutonium to Los Alamos. In case of a wreck or a fire, the men were told to "get the hell out of the road and get upwind," recalled one of the couriers. As Los Alamos got more and more desperate for plutonium, Matthias threw caution to the wind. By the summer of 1945, the plutonium was flying to Santa Fe on C-47 transport planes.

○

ON APRIL 12, 1945, Franklin Roosevelt was sitting in the living room of his retreat in Warm Springs, Georgia, having his portrait painted. He had come there two weeks earlier to escape the strain of the presidency and World War II, which had left him exhausted and frail. Shortly after 1:00 p.m., he complained of a terrific pain in the back of his head and passed out. A doctor arrived and recognized that the president had suffered a massive cerebral hemorrhage. He gave Roosevelt a shot of adrenaline in the heart to try to revive him. But by the middle of the afternoon, the president was dead.

When Eleanor Roosevelt met with Harry Truman that evening to tell him that her husband was dead and that he would now be president, Truman asked if he could do anything for her. She replied by asking if she could do anything for him. "You are the one in trouble now." After being selected as Roosevelt's surprise pick for vice president in 1944, Truman had met with the president just a handful of times, usually in the company of other government officials. Roosevelt hadn't even told him about the Manhattan Project. A few years earlier, as chair of the Committee on Military Affairs in the Senate, Truman had started to investigate the mysterious federal acquisition of property in Washington State. He backed off when Secretary of War Stimson told him that the land was needed for a top-secret military project.

On April 25, Groves and Stimson met with Truman in the Oval Office to brief him on atomic bombs. At this point, Stimson was the grand old man of Washington policymakers. Born in 1867, educated at Phillips Academy, Yale, and Harvard, he was a product of the 19th century who was being forced to deal with the most vexing problem of the 20th. He had been working with Vannevar Bush and James Conant—still the government's most powerful academic advisors through their leadership of the federal Office of Scientific Research and Development—on a plan for the postwar control of nuclear technologies, and the memo he handed to President Truman reflected these discussions. "Within four months," the memo began, "we shall in all probability have completed the most terrible weapon ever known in human history, one bomb of which could destroy a whole city." The United States could not restrict other nations from building atomic bombs, though probably the only nation that could do so soon would be the Soviet Union. The United States therefore needed to consider the international control of nuclear technologies once the existence of atomic bombs became known. The memo called for the establishment of a committee that could recommend actions to prepare for the war's end. US leadership in the development of atomic bombs "has placed a certain moral responsibility upon us which we cannot shirk without very serious responsibility for any disaster to civilization which it would further," Stimson's memo stated.

Groves then handed the president a 23-page memo entitled "Atomic Fission Bombs." Truman objected that he was too busy to read such a long document, but Groves and Stimson insisted, and Truman acquiesced. The memo said that a gun-type bomb using uranium would likely be ready by about August 1, with a second gun-type bomb ready by the end of the year. But an implosion-type bomb would be tested in July and could be used against Japan in August, with new implosion bombs becoming available about every 10 days thereafter. "While the project's primary mission is the development of atomic bombs for use against Japan the tremendous and far reaching implications of the future cannot be and have not been overlooked." Drawing on the thinking of Bush and Conant, the memo concluded: "Atomic energy, if controlled by the major peace-loving nations, should insure the peace of the world for decades to come. If misused it can lead our civilization to annihilation."

In their post-meeting notes, both Stimson and Groves congratulated themselves on how well the meeting had gone. Truman had approved their plans and appeared to be enthusiastic about the new weapons. Now Groves just needed to prove that his $2 billion bombs would work.

A WEEK LATER, Truman established the committee Stimson had requested in his memo. Named the Interim Committee, since new arrangements would be inevitable once atomic bombs became public knowledge, its eight members were all civilians: Stimson was the chair, with George Harrison, his special assistant at the War Department, serving as deputy chair. Undersecretary of the Navy Ralph Bard and Assistant Secretary of State William Clayton—both, like Harrison, former businessmen who had responded to Roosevelt's requests to serve in government—represented two government agencies that would be particularly affected by the development of atomic bombs. Bush, Conant, and Karl Compton, president of the Massachusetts Institute of Technology (and Arthur Compton's brother), represented academia. The final member, serving as President Truman's personal representative, was James Byrnes, a former senator from South Carolina whom Truman would soon name secretary of state. Groves was not on the committee, but he attended all its meetings, providing information and subtly steering the committee in directions he wanted it to take.

At the committee's second meeting on May 14, it decided to create a scientific advisory group that "should be free not only to discuss technical matters but also to present to the Committee their views concerning the political aspects of the problem," as the meeting notes put it. The four members of the panel were Arthur Compton, still at the Met Lab in Chicago; Enrico Fermi, who by this time had relocated to Los Alamos; Ernest Lawrence at the University of California, Berkeley; and the director of the Los Alamos laboratory, Robert Oppenheimer.

Attendance was spotty at most of the Interim Committee's meetings, but all its members were there for the committee's only meeting with the Scientific Panel on May 31, 1945. At 10:00 a.m., Stimson opened the meeting, which took place in a conference room in the Pentagon, by praising "the brilliant and effective assistance rendered to the project by the scientists of the country." The development of nuclear energy marked "a new relationship of man to the universe," Stimson said, that "might be compared to the discoveries of the Copernican theory and of the laws of gravity, but far more important than these in its effects on the lives of men." Then the committee and its Scientific Panel got down to work.

As Oppenheimer explained, the bombs being prepared at Los Alamos would have an explosive force of between 2,000 and 20,000 tons of TNT. The wide uncertainty in the estimate came not from the uranium bomb, which physicists were sure would produce an explosion about in the middle of that range, but from the implosion bomb. Until it could be tested, no one knew whether implosion would produce a relatively small explosion or something much bigger. Beyond their destructive power, Oppenheimer emphasized the psychological and health impacts of nuclear weapons. They would produce a "brilliant luminescence" that would rise two to four miles into the air. In addition, they would shower their surroundings with neutrons "dangerous to life for a radius of at least two-thirds of a mile."

The committee and its scientific advisors began to discuss what would happen once the war was over. Lawrence, who had become more hawkish over the course of the war, insisted that the United States needed to know more and do more than any other country to stay ahead in both nuclear research and the application of that research. He "recommended that a program of plant expansion be vigorously pursued and at the same time a sizable stock pile of bombs and material should be built up," the meeting notes say. Oppenheimer countered with an argument made by many Manhattan Project scientists. Fundamental knowledge of nuclear physics existed throughout the world. Other nations would rapidly catch up with the United States once the existence of atomic bombs became public knowledge. Exchanging information with other nations would enhance international stability and human welfare in a nuclear world. Furthermore, "if we were to offer to exchange information before the bomb was actually used, our moral position would be greatly strengthened."

Others at the meeting wondered whether international control was feasible. How could the United States remain permanently ahead of other nations, and especially the Soviet Union, if it made all its nuclear research public? How could the United States be sure that other countries were not building nuclear weapons no matter what they said publicly? Army Chief of Staff Marshall, who was not a member of the Interim Committee but had been invited by Stimson to attend the meeting, cautioned against putting too much faith in an inspection system to ensure compliance with a system of international controls.

The most powerful objections to sharing information with other countries came from Byrnes, who, as the president's representative, inevitably had more sway than the others. Byrnes, whom Roosevelt had passed over when he unexpectedly picked Truman as his vice president in 1944, was always intensely focused on politics. If nuclear information were provided to other nations, he observed at the meeting, the Soviet Union, as a wartime ally of the United States, would demand the information, too. Bush objected that the most sensitive information could be kept secret. But Byrnes insisted that the United States should keep all its nuclear knowledge secret. As the notes put it, he expressed "the view, *which was generally agreed to by all present* [emphasis in original], that the most desirable program would be to push ahead as fast as possible in production and research to make certain that we stay ahead."

The meeting adjourned at 1:15 for lunch in a Pentagon cafeteria, during which the participants apparently divided themselves among tables. At one table, Byrnes asked Lawrence about something he had heard Lawrence mention earlier in the day—something about demonstrating the bomb's power to the Japanese to convince them to surrender. Yes, said Lawrence, some of the scientists with the Manhattan Project believed strongly that an atomic bomb should be demonstrated before it was dropped on Japanese cities. Others at the table raised objections to the idea. What if the bomb didn't work, since it hadn't been tested? Wouldn't that just strengthen Japanese resolve? If the demonstration were in Japan, military leaders could move American prisoners of war into the area or shoot down the plane delivering the bomb. The American fire-bombings of Japanese cities had already killed more people than an atomic bomb probably would. Wasn't a nuclear weapon just a continuation of bombing by other means?

With the question of a demonstration unresolved, the meeting resumed at 2:15, and the conversation soon turned to potential targets for nuclear weapons. Groves and other military leaders had already been working on targeting, and the Interim Committee had not been asked to examine the issue. Nevertheless, the committee discussed targeting at length, both in the morning and afternoon. The bomb should make as profound a psychological impression on the people of Japan as possible, they said. According to the meeting notes, Conant suggested that "the most desirable target would be a vital war plant employing a large number of workers and closely surrounded by workers' houses." This wording made an impression. In the meeting of the Interim Committee the next day, at which the Scientific Panel was not present, the notes record that "Mr. Byrnes recommended, and the Committee agreed, that the Secretary of War should be advised that, while recognizing that the final selection of the target was essentially a military decision, the present view of the Committee was that the bomb should be used against Japan as soon as possible; that it be used on a war plant surrounded by workers' homes; and that it be used without prior warning."

The power of the Interim Committee to influence the use of atomic bombs has been much debated. To some extent, it was established as a way to quell the objections of scientists who had begun arguing against the use of an atomic bomb in Japan. Groves also had an interest in manipulating the Interim Committee so that it would arrive at conclusions he had already made. By this time, Groves had been planning for months to drop atomic bombs on Japanese cities, and the committee's idea that such a bomb be used on a "war plant surrounded by workers' homes" was a non sequitur if the committee meant that the bomb should not be used on cities. War plants in Japan were in the middle or on the outskirts of cities, as they were in the United States.

The Interim Committee also failed to consider many critical issues. For example, it never appears to have discussed the number of Japanese civilians who might be killed by nuclear bombs. By the end of the war, that issue ranked relatively low in importance. On September 1, 1939, the day Germany invaded Poland to start World War II, President Roosevelt called on every nation engaged in hostilities "publicly to affirm its determination that its armed forces shall in no event, and under no circumstances, undertake the bombardment from the air of civilian populations or of unfortified cities, upon the understanding that these same rules of warfare will be scrupulously observed by all their opponents." Yet beginning with the Japanese bombing of Chinese cities in the 1930s; the attacks by Italian and German fascists on the Spanish cities of Barcelona, Granollers, and Guernica in 1937–1938; Germany's bombing of Warsaw, Rotterdam, London, and other cities during World War II; the widespread bombing of Germany by Britain and the United States; and the United States' incendiary bombing of Japanese cities starting in March 1945, the leading industrial nations of the world pursued a steady descent into barbarism. By the last few months of World War II, some prominent military officers and public figures still recoiled from the use of indiscriminate bombing in war, but their voices were a minority and easily ignored. Especially regarding Japan, the question of civilian casualties was essentially moot. The attack on Pearl Harbor, the wartime atrocities committed by Japanese soldiers, the vicious fighting taking place on islands in the Pacific, and the overt racism of US news reports and public opinion all created a thirst for blood.

If the Interim Committee had never existed, things probably would have turned out more or less the same. Yet its existence also provided an opening for one of the most provocative and far-reaching documents from the dawn of the nuclear age.

AT ITS MEETING with the Interim Committee, the four members of the Scientific Panel had been told that they could present their views to the committee at any time. Arthur Compton returned to Chicago determined to take up the committee's offer. He quickly organized five groups, with a sixth added later, to prepare reports for an upcoming meeting of the Scientific Panel. These groups were mostly organized around issues of postwar research, production, and education. But one—the Committee on Social and Political Implications—had a broader mandate. Chaired by James Franck, its members included Leo Szilard, Glenn Seaborg (who later wrote that Szilard had the greatest influence on the committee), and biophysicist Eugene Rabinowitch, who later cofounded and edited the *Bulletin of the Atomic Scientists*.

The committees were not starting from scratch in preparing their input for the Scientific Panel. For more than a year, the intensity of the research being done at the Met Lab had declined as work picked up at Hanford, Oak Ridge, and Los Alamos. Some of the scientists and engineers had moved to other locations, such as Fermi. Others, like Seaborg, remained in Chicago doing research on the many technical problems associated with nuclear energy. Various groups of Met Lab scientists had already put together reports on the future of what they had come to call nucleonics. Now the thinking that went into these reports fueled the work of the Franck Committee. The week after Compton's return from Washington, DC, Franck and his committee met several times and began writing. Rabinowitch took the lead in drafting the committee's report, with Szilard and Franck suggesting much of the report's content. By the following weekend, what became known as the Franck report was done.

"The scientists on this Project do not presume to speak authoritatively on problems of national and international policy," the report stated in its preamble. "However, we found ourselves, by the force of events, the last five years in the position of a small group of citizens cognizant of a grave danger for the safety of this country as well as for the future of all the other nations, of which the rest of mankind is unaware." The committee pointed out that the development of nuclear power was much more dangerous than any previous invention. It gave humanity the ability to wreak unlimited devastation upon itself, to create "a Pearl Harbor disaster, repeated in thousand-fold magnification, in every one of our major cities." Other nations would quickly catch up with the United States after the war, it observed. In particular, "the experience of Russian scientists in nuclear research is entirely sufficient to enable them to retrace our steps within a few years, even if we make all attempts to conceal them." Nor could the United States hope to monopolize the materials needed to make nuclear weapons. The Soviet Union controlled about one-sixth of the land area of the Earth. Certainly it would be able to find all the uranium it needed.

Because the United States could not monopolize nuclear weapons, a nuclear arms race among nations was sure to ensue, the report said, unless nuclear materials and technologies could be brought under some form of international control. As an option, the report suggested forming a partnership of nations to control and ration the raw materials needed to build weapons—primarily uranium ore. "Efficient controls" then would need to be in place to ensure compliance with international agreements not to divert nuclear materials to weapons. The report did not explicitly call for international inspection of possible weapons-producing plants, but it noted that "no paper agreement can be sufficient, since neither this or any other nation can stake its whole existence on trust into other nations' signature."

Finally, the committee stated that "the way in which nuclear weapons, now secretly developed in this country, will first be revealed to the world appears of great, perhaps fateful importance." If bombs were to be dropped on Japanese cities without warning, the report said, the prospects for international control would be severely damaged.

> Russia, and even allied countries which bear less mistrust of our ways and intentions, as well as neutral countries, will be deeply shocked. It will be very difficult to persuade the world that a nation which was capable of secretly preparing and suddenly releasing a weapon as indiscriminate as the rocket bomb and a million times more destructive is to be trusted in its proclaimed desire of having such weapons abolished by international agreement.

Instead of dropping an atomic bomb on cities, the report argued, "a demonstration of the new weapon may best be made before the eyes of representatives of all the United Nations, on the desert or a barren island. . . . After such a demonstration the weapon could be used against Japan if a sanction of the United Nations (and of public opinion at home) could be obtained, perhaps after a preliminary ultimatum to Japan to surrender or at least to evacuate a certain region as an alternative to the total destruction of this target."

On Monday, June 11, Franck got on the train to deliver the report personally to Stimson in Washington, DC. But he was told that Stimson was out of town—which was not true—so he left the report with Stimson's assistant Harrison. Meanwhile, Compton had written a cover note for the report that essentially undercut its main arguments. The report failed to point out that forgoing use of atomic bombs could make the war longer and cost human lives, Compton's letter stated. He also wrote that if atomic bombs were not used for military purposes, "it may be impossible to impress the world with the need for national sacrifice in order to gain lasting security." In other words, the world had to see what atomic bombs were capable of doing before the public would agree to reining them in.

Whether Stimson ever read the Franck report remains questionable, and he certainly never passed it on to Truman. As the test date for the plutonium bomb drew nearer, he was getting advice about the use of nuclear weapons from many people, both inside and outside government. As historian Alice Smith has written, "In the light of what has transpired the Franck Report strikes many people as a singularly moving and prescient statement; in the busy days of June, 1945, it was one of an endless succession of memoranda to be read if time permitted."

○

THE WEEK AFTER Franck took his committee's report to Washington, DC, the four members of the Scientific Panel—Oppenheimer, Fermi, Lawrence, and Compton—gathered in Los Alamos to prepare their follow-up input to the Interim Committee. They were all familiar with the arguments of the Franck report, even if they had not read its full text. But they had received a request from Stimson's assistant Harrison to comment separately on whether an atomic bomb should be demonstrated first or used without warning on a Japanese city.

It was a dreadful assignment. The Scientific Panel did not know enough about the political, military, and economic circumstances of the war to make an informed recommendation—and they admitted as much in their report to the Interim Committee. But they all had worked on the Manhattan Project, and all had thought carefully about how atomic bombs might be used. In the end, they accepted the Interim Committee's charge.

In a brief report over Oppenheimer's signature, the Scientific Panel said:

> The opinions of our scientific colleagues on the initial use of these weapons are not unanimous: they range from the proposal of a purely technical demonstration to that of the military application best designed to induce surrender. Those who advocate a purely technical demonstration would wish to outlaw the use of atomic weapons, and have feared that if we use the weapons now our position in future negotiations will be prejudiced. Others emphasize the opportunity of saving American lives by immediate military use, and believe that such use will improve the international prospects, in that they are more concerned with the prevention of war than with the elimination of this specific weapon. We find ourselves closer to these latter views; we can propose no technical demonstration likely to bring an end to the war, and we see no acceptable alternative to direct military use.

In later years, all four of the scientists had—or were reported by other members of the panel to have had—misgivings about recommending "direct military use" in their report. But the panel member who reportedly held out longest against the report was Fermi. According to a 1983 interview with Oppenheimer's secretary, Oppenheimer and Fermi argued until five o'clock in the morning before Fermi accepted the statement Oppenheimer had written. Since the beginning of the Manhattan Project, Fermi had become increasingly alarmed by the prospect of nuclear weapons. Once, while visiting Los Alamos, he was shocked by the enthusiasm of the scientists Oppenheimer had recruited to the lab. "I believe your people actually *want* to make a bomb," he had said. If atomic bombs did work, Fermi believed, the United States should keep their existence secret for as a long as possible. That way, the inevitable escalation of tensions created by nuclear capabilities could at least be postponed.

Fermi never expressed regret in later years for signing the Scientific Panel's statement, but Oppenheimer did. "We didn't know beans about the military situation in Japan," he later admitted. "We didn't know whether they could be caused to surrender by other means or whether the invasion was really inevitable. But in the back of our minds was the notion that the invasion was inevitable because we had been told that."

The argument that appears to have persuaded the members of the Scientific Panel is the one Compton used in his letter to Stimson— that the full horror of atomic weapons had to be experienced for the world to turn forcefully against them after the war. If the use of atomic weapons "would result in the shortening of the war and the saving of lives—if it would mean bringing us closer to a time when war would be abandoned as a means of settling international disputes—here must be our hope and our basis for courage," Compton later wrote.

The Interim Committee met again in Washington, DC, on June 21, and toward the end of the meeting it turned its attention to the Scientific Panel's report. The report gave the committee no need to reconsider its former decision. "The committee *reaffirmed* the position taken at the 31 May and 1 June meetings," the meeting notes state, "that the weapon be used against Japan at the earliest opportunity, that it be used without warning, and that it be used on a dual target, namely, a military installation or war plant surrounded by or adjacent to homes or other buildings most susceptible to damage."

○

IN THE FIRST FEW MONTHS of 1945, just as plutonium production was ramping up at Hanford, strange objects began to appear in the sky above Washington and other western states. After the Doolittle raid of April 1942, the Japanese military decided that it needed to retaliate against the US mainland. It could not get bombers all the way across the Pacific, so it opted for an unexpected alternative. The jet stream was not formally discovered until B-29s flew high enough during World War II to encounter its high-speed winds. But Japanese meteorologists had suspected its existence earlier, and weapons designers in Japan realized they could use the jet stream to bomb the United States. They built thousands of balloons filled with hydrogen that bore incendiary bombs and 32 dangling sandbags. When the balloons dropped below a certain altitude, an altimeter caused one of the sandbags to drop off. The bomb designers calculated that by the time the thirty-second sandbag was gone, the jet stream would have blown the balloons across the Pacific. They then would descend, explode when they reached the ground, and set fire to American forests.

On March 10, the reactors at Hanford suddenly shut down when the electrical system went out. A balloon bomb from Japan had become entangled in high-voltage electric lines near the site and had burst into flames. Safety systems prevented the reactors from overheating, and they were quickly restarted. As the "Unusual Incident" report stated, "The line was not seriously damaged as evidenced by the fact that it was restored to service in two minutes."

The US military had asked newspapers not to report on the fire balloons so that the Japanese would not know that their bombs were reaching the United States, which might cause them to launch more. But very few of the balloons caused any damage. The jet stream blows strongly enough for the balloons to reach America only in the winter, when American forests are cold, wet, and unlikely to ignite from a firebomb. By the spring of 1945, the Japanese had given up on the campaign.

On May 5, 1945, minister Archie Mitchell from Bly, Oregon, and his pregnant wife Elsie took five Sunday school children on a picnic in the woods near their town. While the minister was unloading the car, his wife and the children went to investigate a mysterious object they had found nearby. When they moved it, the balloon bomb exploded, and all six were killed. Today, a monument at the site, erected by the Weyerhaeuser Company, reads, "Only Place on the American Continent Where Death Resulted from Enemy Action During WW II."

Chapter 14

TRINITY

IN THE LAST FEW DAYS OF JUNE 1945, METALLURGIST CYRIL SMITH
and his colleagues in the Chemistry and Metallurgy Division at Los
Alamos extracted from a hot-press mold two solid hemispheres of
plutonium. Combined, they were about the size of an apple—a bit
more than three-and-a-half inches across. They would fit "rather easily, and even rather pleasantly, in the hand," Smith later recounted,
though they were incongruously heavy—about the weight of a bowl-
ing ball. Pure plutonium metal has a steel-gray color when it is cast,
but it tarnishes quickly in air. To keep that from happening, and
to avoid contaminating anyone who handled the hemispheres, the
metallurgists had plated the plutonium with a layer of nickel. But
the flat surface of one of the hemispheres had blistered, which was
a potential disaster. Any imperfection in the surface could cause
asymmetries during implosion, which might cause the bomb to fail.
Smith's team came up with a solution. They placed a layer of gold
foil between the two hemispheres, which they hoped would render
the compression uniform. "My fingers were the last to touch those
portentous bits of warm metal," Smith recalled many years later.
"The feeling remains with me to this day."

The other pieces of the world's first atomic bomb were coming
together at the same time. The best explosive lenses emerging from
the candy-making machines were packed into a five-foot sphere of
an aluminum-copper alloy known as Duralumin, which had been
used in previous decades to make the frames of airships. The det-

onation system, after a frustrating spring spent trying to get the 32 detonators to fire simultaneously, was finally working, though the director of the test suggested stocking up on rabbits' feet and four-leaf clovers. The design of the walnut-shaped initiator, which would fit into a hollow in the center of the plutonium and produce neutrons at the height of implosion, was finalized at the end of April, and the first complete one was manufactured in June. A technician accidentally dropped that first initiator and watched with horror as it rolled toward a drainpipe. It stopped just short.

Meanwhile, a new and highly unusual laboratory was taking shape on a barren plain 150 miles south of Los Alamos. The region was known as the Jornada del Muerto—the dead man's journey—a name given it by conquistadors trudging across the desolate flatlands from New Spain. About two miles from a four-room ranch house appropriated by the army at the beginning of the war, metalworkers erected a surplus Forest Service fire watchtower a hundred feet tall. Ten miles south of the tower, a straggly basecamp housed the 250 or so people working at the site. About six miles north, south, and east of the tower, concrete shelters dug into the earth were filled with electrical equipment, cameras, radiation detectors, and searchlights. When the time came to implode the plutonium in the test device atop the tower, observers in the shelters could measure the bomb's effects.

Oppenheimer had code-named the test Trinity. He later cited several sources as inspirations for the name, including part of a John Donne poem he admired:

Batter my heart, three person'd God; for, you
As yet but knocke, breathe, shine, and seeke to mend;
That I may rise, and stand, o'erthrow mee, and bend
Your force, to breake, blowe, burn and make me new.

THROUGHOUT THE SPRING, as German opposition collapsed, Groves had been thinking about where to drop atomic bombs on Japan. As the Trinity test approached, the work to identify targets intensified.

Groves had enlisted a committee of assistants, military men, and technical experts from Los Alamos to help him sort through the choices. Known as the Target Committee, the group held its first meeting on April 27 at the Pentagon and began laying out the criteria that potential targets must meet. First, the men dropping the bombs should be able to see the target rather than relying on radar. That way, the likelihood of a direct hit would be substantially increased and observers on nearby planes could document the bombs' destructiveness. At the committee's first meeting, this criterion led to an extensive discussion of late-summer weather in Japan, which tends to be cloudy and hot. Only a few days per month were likely to have conditions suitable for visual bombing. Decisions on exactly when and where to drop atomic bombs would therefore have to be left to commanders in the field. Only they would have access to all the information, including weather information, needed to use the bombs effectively.

Second, the bombs should be used in such a way as to have the greatest possible impact on the Japanese psyche. Though Japan was clearly headed toward defeat in the spring of 1945, brutal combat was still under way on Okinawa, and hardliners in the Japanese government vowed that they would keep fighting even if the Americans launched an invasion of the Japanese mainland. Many US military and civilian leaders thought that the Japanese needed to be subjected to some sort of psychological shock to convince them to end the war, and using atomic bombs against Japanese cities seemed perfectly suited to the task. Granted, the cities chosen should have military installations, including either "important headquarters or troop concentrations, or centers of production of military equipment and supplies," as Groves put it in his memoirs. But most Japanese cities contained such installations. Furthermore, as the Target Committee concluded at its third and final meeting on May 28, the bombing crews should "endeavor to place first gadget in center of selected city" rather than relying on subsequent bombs to finish the job. Small and isolated military targets in remote locations would be hard to hit and would not provide the necessary shock value. Only cities would do.

Third, the targets should be relatively undamaged before the bombing so that the full effects of the bombs could be determined. This was getting increasingly difficult by the summer of 1945. In March, the air forces began bombing Japanese cities from air bases in the recently captured Mariana Islands with the intention, as expressed in the notes of the first Target Committee meeting, "of not leaving one stone lying on another." Eventually, more than 60 cities in Japan were attacked and hundreds of thousands of civilians were killed, including nearly a hundred thousand people in the March 9–10 firebombing of Tokyo. Groves and Stimson eventually had to plead with the leaders of the air forces to leave a handful of cities untouched.

One city perfectly met all the committee's criteria, as far as Groves was concerned: Kyoto, the ancient capital of Japan. With a population of about a million, Kyoto was Japan's fourth largest city. It was also an intellectual and religious center, home to thousands of temples and shrines. Destroying these cultural treasures in a single fearsome blow was just the kind of psychological shock Groves hoped to administer. Moreover, as the Target Committee notes put it, "Kyoto has the advantage of the people being more highly intelligent and hence better able to appreciate the significance of the weapon." In the list prepared by the Target Committee in May, Kyoto was on top, followed by four more cities: Hiroshima ("an important army depot and port of embarkation in the middle of an urban industrial area"), Yokohama ("an important industrial area which has so far been untouched"), Kokura Arsenal ("one of the largest arsenals in Japan . . . surrounded by urban industrial structures"), and Niigata ("its importance is increasing as other ports are damaged").

Well before the Target Committee began its work, Groves had

become convinced that two atomic bombs would be needed to force Japan to surrender, in part because of a conversation he had had with Navy Rear Admiral William Purnell. The first would show that the United States could build such weapons, Purnell told him. The second would demonstrate that the United States could keep building atomic bombs until the Japanese surrendered.

But of course Groves had another reason for dropping two bombs on Japan. He had spent $2 billion on two different approaches to nuclear weapons: one using uranium from Tennessee, the other using plutonium from Hanford. Dropping two bombs would justify both.

On May 30, two days after the third and final Target Committee meeting, Groves drove across the Potomac River to meet Stimson in the Pentagon. During the meeting, according to Groves's recollections, Stimson asked whether he had selected the targets to be bombed. Groves replied that a report on potential targets had been completed and that he was planning to bring it to General Marshall the next day. Stimson asked to see it. Groves replied, "It's across the river and it would take a long time to get it."

Stimson said, "I have all day and I know how fast your office operates. Here's a phone on this desk. You pick it up and you call your office and have them bring that report over." When Groves continued to object that Marshall should see the report first, Stimson replied, "This is one time I'm going to be the final deciding authority. Nobody's going to tell me what to do on this. On this matter I am the kingpin and you might just as well get that report over here."

As soon as the report arrived, Stimson read through it. He demanded that Kyoto be removed from the list. He said that it was the cultural center of Japan and he would not see it destroyed. Stimson then rose from his chair, walked to General Marshall's adjacent office, and asked if he was busy. "Groves has just brought me his report on the proposed targets," he told Marshall. "I don't like it. I don't like the use of Kyoto."

Opinions differ on exactly why Stimson was so opposed to bombing Kyoto. Stimson had spent several days in Kyoto in 1926 with his wife, and undoubtedly the two of them had visited some of the city's cultural sites; perhaps he was remembering these places as he read Groves's target list. Stimson was also the product of an older, Victorian morality. He often lamented the indiscriminate bombing of civilians during World War II, even as he permitted such bombing to continue. Maybe he used the protection of Kyoto as a way to ease his conscience after so many other Japanese cities had been destroyed. Or he might have been thinking about the need to rebuild relations with Japan after the war. As he wrote in his diary about the possibility of bombing Kyoto, "the bitterness which would be caused by such a wanton act might make it impossible during the long post-war period to reconcile the Japanese to us in that area rather than to the Russians." All these factors could have been at work, but one thing is certain: in demanding that Kyoto be removed from the target list, Stimson prevented an even greater tragedy than the ones that were on the way. With a population of about a million, Kyoto was more than three times the size of Hiroshima and four times the size of Nagasaki. If it had been successfully bombed, the death toll could easily have approached a half million.

For the next two months, Groves tried as many as a dozen times to have Kyoto placed back on the list, but each time Stimson rebuffed him. Finally, Stimson took his case all the way to the president. "As to the matter of the special target [Kyoto] which I had refused to permit," Stimson wrote Groves in July from the Potsdam Conference in Germany, where Truman, Churchill, and Stalin were negotiating over the future of the postwar world, "he [Truman] strongly confirmed my view and said he felt the same way."

Nagasaki was a very late addition to the target list. It was among a list of 17 cities cited as potential targets in the Target Committee's first meeting, but after that it was not mentioned again in the committee's meeting notes. However, Yokohama had been severely bombed by the air forces and was therefore no longer a good target. Niigata, a port on the west coast of Japan north of Tokyo, had taken the place of Kyoto. That left one open slot on the list of four cities prepared by the Target Committee.

On July 24 a message arrived in Washington, DC, from air forces chief Henry Arnold, who was with the president at the Potsdam Conference in Germany. It said that Nagasaki should be added to the list of target cities. The generals in Washington who received the message were not happy with the choice. Nagasaki did not meet the criteria established by the Target Committee. It nestled within two valleys that would contain and redirect the force of the bomb, making it hard to determine the bomb's effects. It also had already been bombed several times, though the damage done was relatively minor. And Nagasaki was unusual among Japanese cities, though no one seems to have acknowledged its history when it was being considered as a target. Nagasaki was a major port for Portuguese and Dutch traders after European contact in the 16th century. When Japan closed its border to foreigners in the first half of the 1600s, it allowed Dutch traders to continue to do business on a small island in Nagasaki's harbor. As a result of this history, Nagasaki remained the most Western of Japanese cities, with a sizable Roman Catholic population and a large cathedral in the city's Urakami Valley. Even after Japan reopened its borders to foreigners in 1854, Nagasaki remained a center of Western learning.

Cables to Potsdam from the generals in Washington, DC, who objected to Nagasaki failed to alter the decision. By the evening of July 24, Nagasaki was the fourth city on the target list.

A July 25 directive, drafted earlier by Groves and revised once the target list was final, laid out the plan:

1. The 509th Composite Group, Twentieth Air Force, will deliver its first special bomb as soon as weather will permit visual bombing after about 3 August 1945 on one of the targets: Hiroshima, Kokura, Niigata and Nagasaki. . . .
2. Additional bombs will be delivered on the above targets as

soon as made ready by the project staff. Further instructions will be issued concerning targets other than those listed above.

This was the only written command ever issued to use atomic bombs on Japan. It was signed not by Truman, who for the most part maintained his distance from decision making about the bombs, but by General Thomas Handy, who was the army's acting chief of staff in Washington, DC, with Marshall away at the Potsdam Conference. As Groves reflected after the war, President Truman did not have to issue an order to use atomic bombs on Japan. The endgame had been preordained long before, when Roosevelt decided to back the Manhattan Project. Groves was only slightly exaggerating when he said after the war that Truman "was like a boy on a toboggan." The technological and bureaucratic momentum of the project was immense, and Groves had a powerful self-interest in keeping that momentum going. He had devoted the last three years of his life to this project and was going to see it through. The only thing that could have stopped the use of atomic bombs was a decision by Truman or a group of his top advisors not to use them—and that was not going to happen. Truman and his advisors had already accepted the idea of killing hundreds of thousands of Japanese civilians with firebombs. Meanwhile, they were playing a delicate diplomatic game with Stalin. At the Yalta Conference earlier that year, Stalin had promised to enter the war against Japan three months after Germany was defeated, which meant the Soviets would attack Japan no later than August 8. Now the use of atomic bombs looked like a way of forcing the Japanese to surrender before the Soviet Union entered the war, thereby preventing the Soviets from demanding territorial or political concessions afterwards.

On the domestic front, Truman would have been politically crucified if he prolonged the war by refusing to use weapons that had cost $2 billion to build. Furthermore, he never seemed to have fully realized—or refused to fully realize—that the bombs were going to be dropped on cities. He seems to have thought that Hiroshima, at least, was a purely military target, even though a single question to one of his advisors would have revealed otherwise. Only after the event did he acknowledge what the bombs had done.

●

LEO SZILARD DIDN'T GIVE UP when the Scientific Panel failed to endorse the Franck committee's call for a demonstration of an atomic bomb. He decided that he would have to act on his own without the other members of the committee.

At the beginning of July, Szilard drafted a petition to President Truman. The development of atomic bombs "places in your hands, as Commander-in-Chief, the fateful decision whether or not to sanction the use of such bombs in the present phase of the war against Japan," the final version of the petition stated. Gone was any mention of demonstrating the bomb. Instead, Szilard wanted the president simply to think hard about the forthcoming decision. The United States should make public the terms it planned to impose on Japan after the war, the petition stated. Once these terms were announced, Japan should be given an opportunity to surrender. "If Japan still refused to surrender," the petition went on, "our nation might then, in certain circumstances, find itself forced to resort to the use of atomic bombs. Such a step, however, ought not to be made at any time without seriously considering the moral responsibilities which are involved." These moral responsibilities were paramount, the petition observed: "A nation which sets the precedent of using these newly liberated forces of nature for purposes of destruction may have to bear the responsibility of opening the door to an era of devastation on an unimaginable scale."

Sixty-nine people at the Met Lab signed the petition—an act of great bravery, given the government's control over their current and future careers. Lacking a friend or acquaintance who could deliver a document to the president, Szilard gave the signed petition to Compton, who in turn gave it to Kenneth Nichols, Matthias's counterpart at Oak Ridge. Nichols then sent the petition to Groves.

Meanwhile, Szilard arranged for several people he knew at Los Alamos to distribute the petition there, where it quickly came to the attention of Oppenheimer. Despite his own involvement in political decision making, the lab director was annoyed that Szilard was trying to influence policy, and he banned drafts of the petition from being distributed in Los Alamos. He also informed Groves about the petition, so that Groves knew what was coming.

Groves had no reason to act on the petition right away, though he knew he would have to deal with it at some point. He had already been discussing with the Interim Committee ways of eliminating "certain scientists of doubtful discretion and uncertain loyalty"—by whom he meant Szilard—from the Manhattan Project. But Groves knew that immediately firing or imprisoning Szilard would trigger a revolt by other scientists that he could not afford at this critical juncture in the project. He would bide his time.

Not until August 1 did Groves forward Szilard's memo to Stimson's office. By then Stimson was caught up in other affairs and gave the petition no notice. It sat unread among the papers on his desk.

●

ON FRIDAY, JULY 13—a date chosen by George Kistiakowsky in the hope that it would bring the project good luck—the scientists and engineers at the Trinity test site south of Los Alamos began the final assembly of the test device.* That afternoon, shortly after 3:00, working beneath the tower in the Jornada del Muerto desert, they removed a brass plug from the center of the assembly. At the very center of the high explosives was a cylindrical void about the size of a small fire extinguisher. The men assembling the bomb readied a

---

* Technically it was not a bomb, since it was not configured for that use, but it was essentially the same device used later as a bomb.

cylinder of the same size that contained Hanford's plutonium. But as they lowered the cylinder into the center of the bomb, it got stuck. Something had gone wrong.

The men thought for a while. Then one of the physicists realized that the bomb had been sitting in the shade beneath the tower while the assembly containing the plutonium had been exposed to the desert heat, which had caused it to expand. "Let it stick there for a few minutes and the heat will be conducted away," he said. Less than a minute later the assembly fell into place—the crisis was over. The team placed the final explosive lenses into the bomb, covering the now-filled hole at its center, and sealed the bomb up.

The next morning they attached a metal cable to a U-shaped bracket holding the bomb, lifted the bomb a few feet into the air, and stacked a pile of mattresses beneath it. If the cable had broken, the mattresses would not have protected the bomb from damage. Still, they made the men feel better.

A hoist raised the bomb to a wooden platform at the top of the tower. An electronics team then climbed the tower to install electrical cables, monitoring devices, and other equipment. It was the afternoon of Saturday, July 14, by the time they had finished. After one final test of the connections, the electrical leads to the bomb were disconnected. They would be reconnected to the five thousand pounds of high explosives and 13 pounds of plutonium when the test was ready to go.

That weekend, various Manhattan Project dignitaries began to arrive in New Mexico. Groves flew to Albuquerque with Bush after the two of them had toured various project facilities, including Hanford. There they met Conant, and the three of them drove south to the Trinity base camp. Three of the four members of the Scientific Panel attended the test. Since the previous fall, Fermi had been living and working at Los Alamos, where he had proved to be a calming influence on the younger scientists. Lawrence had traveled with Groves to New Mexico after joining him on the last part of his West Coast swing. But Compton, whose National Academy of Sciences

The world's first atomic bomb was detonated atop this tower in central New Mexico. *Courtesy Palace of the Governors Photo Archives (NNHM/ DCA), 147362.*

committee had helped set the Manhattan Project in motion, stayed in Chicago. "I could not absent myself at that time without giving rise to questions," he later wrote. Seaborg was not invited from Chicago to attend the first test of the element he had discovered. He wrote in his journal about the test only after it occurred.

James Chadwick, the discoverer of the neutron and the head of the British Mission to the Manhattan Project, had flown to New Mexico from Washington, DC. William Laurence, a science writer at the *New York Times* whom Groves had hired to write about the project, was upset at having to watch the test from 20 miles away rather than from the south bunker, where the shot was being controlled. He had already written his own obituary in case the explosion was larger than anticipated.

The dignitaries arrived not to the hot dry weather they might have expected but to rain, wind, and thunderstorms. A monsoon pattern had set in Saturday that promised to persist for at least two days. By late Sunday night, heavy rains and high winds were lashing the test site, and many of the scientists were convinced that the test, which had originally been scheduled for 2:00 a.m. Monday morning, had to be postponed. Oppenheimer, who had contracted chicken pox a few weeks earlier and now weighed just 115 pounds, was a physical and emotional wreck. He smoked incessantly as he and Groves discussed the weather and the timing of the test. "Groves stayed with the Director, walking with him and steadying his tense excitement," wrote Groves's assistant, Thomas Farrell. "Every time the Director would be about to explode because of some untoward happening, General Groves would take him off and walk with him in the rain, counseling with him and reassuring him that everything would be all right."

Groves was adamant that the test go ahead. He had promised Stimson and other members of the president's delegation in Germany, who were beginning the Potsdam Conference with Churchill and Stalin even as he and Oppenheimer were talking that night, that he would conduct the test as soon as possible. That way, Truman would know whether he had the atomic bomb in his pocket while negotiating with Stalin about ending the war. If the test were delayed, it would take at least several days to re-create the fevered pitch of activity required of the Los Alamos test group, and that would be too late for Truman. Also, Groves was fearful that a saboteur somewhere among the people at Los Alamos was waiting until the last possible minute to disrupt the test. Already he had Manhattan Project scientists stationed at the base of the tower and on the platform with the now fully armed bomb to make sure that no one meddled with it. A delay would just give the saboteur more time to act.

At about 11:00 p.m. Sunday night, Groves urged Oppenheimer to get a couple of hours of sleep before the test, which by then had been rescheduled for 4:00 a.m. Groves drove back to the base camp 10 miles south of the tower, where he was sharing a tent with Bush and Conant. The tent had been poorly staked, which irritated Groves, and Bush and Conant did not sleep a wink with the canvas flapping in the wind. But Groves fell asleep within minutes and slept soundly for about an hour.

At 1:00 a.m., Groves was back with Oppenheimer at the control bunker. By two o'clock the weather forecast had improved slightly, and Groves decided that the test would occur at 5:30 a.m. rather than 4:00. As he later wrote, "There was only one dissenting vote that could have called off the test and that was my own. This operation was not run like a faculty meeting. Advice was sought and carefully considered but then decisions were made by those responsible. There was no one but myself to carry this responsibility."

At 5:10 Monday morning, as the night sky continued to clear, Groves returned to base camp, leaving Farrell with Oppenheimer. Farrell was Groves's understudy on the Manhattan Project. If anything happened to Groves, Farrell was supposed to know enough to step in and finish the job. They had agreed that they should not be in the same place in dangerous situations.

Forty-five seconds before the test was scheduled to occur, Sam Allison, who had been building piles in Chicago before Fermi arrived from New York, began calling out over an intercom the number of seconds left until the test. "Fifty-five, fifty-four, fifty-three, . . ." he said. Allison later said that it was the first time he had ever heard of someone counting backward.

At 5:29:45, Allison reached zero. "Now!" he shouted. Inside the bomb, 32 detonators fired simultaneously atop wedge-shaped pieces of high explosives. The shock wave from the explosives traveled inward and compressed the apple-sized sphere of plutonium at the bomb's core to a bit less than half its original volume—from a diameter of 3.6 inches to 2.6 inches, which is about the size of a tennis ball. At that instant, the initiator in the middle of the plutonium loosed a flood of neutrons into the supercritical plutonium. In the next millionth of a second, more than a million-million-million-million atoms fissioned, each of them painstakingly forged in Hanford's nuclear furnaces.

Several hundred people on the Jornada del Muerto witnessed the explosion of the world's first nuclear bomb that Monday morning. What many of them remembered best was the light. When an atomic bomb explodes, it emits an incredibly bright flash of white light, many times brighter than the sun, the brightest light ever created on Earth, a light bright enough to be easily seen from the moon and nearby planets. Those who were looking away from the tower saw the desert and surrounding mountains lit with a vivid intensity. Every stone, mesquite bush, and crevasse stood out as if illuminated by a flashbulb. Those who were looking in the direction of the tower without the dark welder's lenses they had been issued were temporarily blinded. They missed most of what happened next.

As the bright flash of light faded, a huge white fireball, a thousand feet across, appeared on the horizon at the vaporized tower's former location. Dust kicked up from the ground momentarily obscured the bottom half of the fireball. But then the fireball began to rise, a roiling, bulbous, fearsome sphere of light and gas and dust. After a few seconds the white light began to fade and change color, to red, orange, and purple. A thin stem extended from the rising fireball to the ground, connecting the monstrous sphere to its earthly origins. The churning fireball rose higher and higher, above a layer of clouds, gradually darkening in the predawn sky. As George Kistiakowsky, who watched the test with Oppenheimer at the control bunker, later reflected, "at the end of the world—in the last millisecond of the Earth's existence—the last men will see what we saw."

The tremendous heat of a nuclear explosion compresses the air around the bomb, which creates a shock wave that propagates in all directions. This shock wave was so powerful that it knocked Kistiakowsky down six miles away from the blast. The sound of the explosion reached the observers as a loud crack. The sound then began reverberating off the surrounding mountains. For several minutes the Earth roared as the fireball rose higher into the sky.

At the base camp, Fermi was so focused on an experiment he was doing that he paid little attention to the light or sound from the explosion. He waited about 40 seconds for the shock wave to travel the 10 miles from the explosion. He then began dropping scraps of paper from his hand. He measured the distance they were blown sideways with his shoe. From this measurement, he calculated that the yield of the blast was equivalent to 10,000 tons of TNT. His estimate was off by only about a factor of two.

An hour or so after the test, as the sun began to light the desert, Oppenheimer and Farrell drove back to base camp. When the two of them met Groves, Farrell said, "The war is over."

"Yes," Groves replied, "after we drop two bombs on Japan."

Chapter 15

TINIAN ISLAND

CHARLES SWEENEY EASED OFF THE BRAKE, AND THE OVERLOADED B-29, its four 18-cylinder engines roaring, lurched down the runway. Someone had turned off the lights at the runway's far end, though Sweeney didn't know why. He would have to assume that he could get his plane in the air before the macadam gave way to the Pacific Ocean. The previous evening he had watched a heavily loaded B-29 crash on takeoff, its napalm-filled bombs and fuel tanks bursting into flames. But Sweeney's flight was different. If his plane crashed, the 13 pounds of Hanford plutonium it carried could vaporize much of the air base on Tinian Island.

Sweeney desperately wanted this mission to go perfectly. Three days earlier, on August 6, his friend and commanding officer Paul Tibbets had flown a different B-29, the *Enola Gay*, to drop the uranium-based bomb on Hiroshima, and that mission had been flawless. The weather had been ideal, Tibbets had dropped the bomb on his first run, and much of the city had been destroyed. Sweeney had also been flying a B-29 over Hiroshima on that mission, *The Great Artiste*, which was filled with instruments to measure the effects of the blast. Looking back on the city, Sweeney had seen Hiroshima covered with a roiling brown cloud interspersed with flames. Above the city, and far above his plane, rose a thin vertical column of smoke filled with colors, colors Sweeney had never seen before. It was a beautiful and terrible thing to see.

As Sweeney had rolled *The Great Artiste* to a stop back on Tinian

Island, he had glimpsed, through the plexiglass window of the B-29, the commander of the army air forces pinning the Distinguished Service Cross on Tibbets's flying coveralls. Sweeney was proud to be associated with the commander of the 509th Composite Group. Tibbets had been leading the group for less than a year, but in that time he had fashioned it into a tight-knit, efficient operation. Just 30 years old, reserved, and soft-spoken, Tibbets had the assurance of a man many years older. Sweeney had wanted to join his outfit from the first day they met.

The evening of the Hiroshima bombing, amidst a raucous celebration party, Tibbets had pulled Sweeney aside. What Tibbets had to say shocked Sweeney. Tibbets wanted the younger man—Sweeney was just 25—to command the mission to drop a second atomic bomb on Japan. Sweeney had always assumed that Tibbets would lead all the missions. Now Tibbets was turning the job over to him. As Sweeney later said, "I'd rather face the Japanese than Tibbets in shame if I made a stupid mistake."

Accelerating down the runway, Sweeney glanced at the *Bockscar*'s speedometer—125 miles an hour. He kept the yoke pushed forward, the wheels down. The B-29, with its 10 regular crew members, three additional men to monitor the 10,300-pound bomb, and 7,250 gallons of fuel, was way above Boeing's weight guidelines. Sweeney would have to use the entire runway to get the plane aloft. He glanced again at the speedometer—140 miles an hour. In the darkness, he could feel rather than see the water approaching. At 155 miles an hour he eased up on the yoke. The plane rose slowly from the end of the runway and soared out over the Pacific Ocean.

Three days earlier, for the Hiroshima mission, the weather had been calm and clear. Now the skies were filled with clouds and distant flashes of lightning—a bad omen, Sweeney thought. Already this mission had encountered more problems than the entire mission to Hiroshima. During preflight checks, Sweeney had discovered that a pump connected to two 320-gallon reserve fuel tanks in the rear bomb bay was not working. He had gotten off the plane to discuss

Technicians on Tinian Island prepare the bomb, nicknamed Fat Man, that was dropped on Nagasaki on August 9, 1945. *Courtesy of the US National Archives and Records Administration.*

the situation with Tibbets. Replacing the fuel pump would take at least several hours, and the bomb could not be moved to another plane. If they waited until the pump was fixed, the mission would likely have to be postponed for a day. But the mission had already been moved up two days, to Thursday, August 9, because the weather was supposed to turn stormy on Friday and stay bad for at least five days. Sweeney knew that even without the fuel in the reserve tank, he should have enough to fly to Japan, drop the bomb, and return to Tinian Island. Tibbets had said, "It's your call, Chuck."

"The hell with it. I want to go."

The *Bockscar* had no choice but to fly through the thunderstorms on the way to Japan. The plane yo-yoed through the sky, the crew bucking within their four-point restraints. Lightning backlit the clouds; St. Elmo's fire played along the props and wingtips. Sweeney, his copilot, and a third pilot on the flight tried not to jostle the bomb hanging from a shackle in the forward bomb bay of the plane, but there was nothing they could do. The bomb was egg-shaped, about 10 feet long and five feet around, with a set of boxy fins attached to one end. It was painted yellow, like the practice bombs they had been dropping for months, with black sealant spray-painted on its metal seams. The tail fins of the bomb were covered with signatures and messages to the Japanese from the Tinian ground crew. "A second kiss for Hirohito," wrote Rear Admiral Purnell, whose arguments had convinced Groves that two bombs would be necessary to defeat Japan.

After the Hiroshima bomb had been dropped, President Truman had said in a statement, "We are now prepared to obliterate more rapidly and completely every productive enterprise the Japanese have above ground in any city. We shall destroy their docks, their factories, and their communications. Let there be no mistake; we shall completely destroy Japan's power to make war. If they do not now accept our terms they may expect a rain of ruin from the air, the likes of which has never been seen on this earth." But Japan had not accepted the US terms. Japanese radio stations were already downplaying the extent of Hiroshima's destruction. So many of Japan's cities had already been firebombed that the nation's leaders seemed not to care much about one more being destroyed. They seemed to be more concerned about the Soviet Union's entry into the war the previous day. Soviet tanks rolling into Manchuria demonstrated once and for all that Japan could not rely on any other country to come to its rescue. Yet the Japanese leaders still refused to surrender.

Sweeney's mission was to drop the second bomb on Japan as quickly as possible to show that the United States would keep punishing the Japanese until they gave up. But Sweeney knew that the United States did not have a ready supply of atomic bombs. The next bomb, after the one hanging in the bomb bay of the *Bockscar*, would not be ready for several weeks. If he failed in his mission, Truman's threat would be empty. Sweeney also knew that the top brass considered this bomb even more important than the first. During the previous evening's briefing, Tibbets had said that this bomb made the Hiroshima bomb obsolete. All future bombs would be based on this design, which meant that the scientists and military leaders were especially interested in what it could do. That's why two other B-29s were on this mission, too—to document the bomb's destructiveness.

Suddenly, as if walking through a doorway, the *Bockscar* emerged from the storm. The crew members collectively exhaled. A few settled into their seats to try to get some sleep. Through the window above his head, Sweeney could see a few stars amidst the remnant clouds. A faint band of dawn appeared on the eastern horizon.

On the flight deck behind Sweeney, Fred Ashworth and his assistant Phil Barnes were monitoring a box covered with switches and lights that would alert them if there was a problem with the bomb's fuses or firing circuits. Ashworth and Barnes had been added to Sweeney's crew to ready and monitor the bomb on the way to Nagasaki. Exactly who was in charge of dropping the bomb was never made clear to either Sweeney or Ashworth, which would cause problems later in the mission. But even before then, Ashworth and Barnes had to endure a crisis. Moving the mission up two days because of the weather had played havoc with the ground crew's schedule. The crew had worked around the clock to get the bomb assembled and loaded onto the *Bockscar*. Inevitably, mistakes occurred.

Not long after emerging from the storm, Barnes woke Ashworth from a nap. "We got something wrong here," Barnes said. "We got a red light going off like the bomb is going to explode right now."

A light on the side of the flight test box was blinking furiously.

"Oh my God," said Ashworth. "Do you have the blueprints? This bomb can pre-detonate if we drop below a predetermined level. What's our altitude? Where are the blueprints?"

With the blueprints rolled out beside them, Barnes and Ashworth removed the casing from the test box and began going over the wiring. Ten desperate minutes later they discovered the problem. The settings of two switches had been reversed. Barnes flipped the switches to their proper positions, and the red light went back to normal.

After about four hours of flying, Sweeney spotted the small circular island of Yakushima, about 40 miles south of Japan's main islands. Within a few minutes he caught sight of the instrument plane, which was being flown by Fred Bock, the usual pilot of the eponymous *Bockscar*. As with the Hiroshima mission, three planes were supposed to travel together: one to drop the bomb, one to measure its effects, and one to photograph the explosion. When Tibbets had chosen Sweeney to lead the mission, Sweeney and Bock had switched planes, since there was not enough time to move the observation instruments from *The Great Artiste* to the *Bockscar*. But neither Sweeney nor Bock could see the third plane that was supposed to meet them at the rendezvous. That plane, the *Big Stink*, was being flown by James Hopkins, a 26-year-old Texan. Sweeney and Hopkins had a strained relationship. They had bickered the previous day over the proper way to conduct the rendezvous over Yakushima. "I know all about that," Sweeney later reported Hopkins saying. "You don't have to tell me how to make a rendezvous." Now their spat was having dire consequences. Hopkins, in the *Big Stink*, was flying several thousand feet above Sweeney in long doglegs, not in the tight circles Sweeney was following. "Where the hell is he?" Sweeney shouted to his copilot.

At the previous evening's briefing, Tibbets had told Sweeney and his crew that they were to wait only 15 minutes at the rendezvous site. If one of the other planes did not show up in that time, they were to leave immediately for the target. But Sweeney did not want to leave. Hopkins was flying the photography plane, which was not essential to the mission, but not having Hopkins there would make the mission incomplete. Despite his orders, Sweeney circled over Yakushima for 45 minutes. During this time, weather reports arrived from two other B-29s that had flown ahead to the primary target, Kokura Arsenal, and the secondary target, Nagasaki. The weather over the primary target was clear, reported the *Enola Gay*. The *Laggin' Dragon* said that the weather over the secondary target was almost as good—only two-tenths cloud coverage.

"The hell with it," Sweeney finally said. "We can't wait any longer." He wiggled his wings at Bock, and the two planes swung away to the north.

Hopkins, flying his lazy doglegs high above, never saw the two planes below him. All the planes were under strict orders not to use their radios, but Hopkins, increasingly frantic, decided that he had to take the risk. "Has Sweeney aborted?" he radioed back to Tinian. But the first word was lost in transmission, and the radio operators on Tinian only heard "Sweeney aborted." When Farrell, who had come to Tinian to serve as Groves's representative on the island, heard that message, he ran outside and threw up.

Kokura Arsenal was just to the south of the spot where two of Japan's four major islands—Kyushu to the south and Honshu to the north—come within a few miles of each other. It was sure to be heavily guarded. The arsenal was a major source of munitions for the army and was surrounded by heavy industry and workers' homes. Sweeney also knew that the Japanese, after Hiroshima, would be on the lookout for three-plane formations flying over unbombed cities. Even though the *Big Stink* had gone missing, he felt highly visible.

At 9:45 local time (an hour earlier than Tinian time), after almost seven hours of flying, the *Bockscar* approached Kokura. The delay over Yakushima had been costly. During the time Sweeney had been looking for Hopkins and flying to the target, clouds had rolled in from the ocean. Also, since the US bombing had not stopped after the destruction of Hiroshima, more than 200 B-29s had firebombed the nearby industrial town of Yawata the day before, and smoke from the still-burning city had drifted east over Kokura. Workers at the arsenal may even have been burning oil in drums or emitting steam to create a smokescreen, though those reports proved difficult to verify in future years.

Despite the clouds and haze, Sweeney lined up the *Bockscar* for a bombing run. Below and in front of Sweeney, in the plane's plexiglass nosecone, bombardier Kermit Beahan readied his Norden bombsight to pick out the target. Beahan, who turned 27 that day, was a friendly and dapper Texan who was widely considered one of the best bombardiers in the army. Sweeney's usual plane, *The Great Artiste*, had been named for Beahan, reflecting his skills as both a bombardier and a ladies' man. But Beahan could see nothing through the clouds. "I can't see the goddamned target," Beahan said over the plane's intercom. "There's a goddamned cloud over the goddamned target."

"No drop," Sweeney said. "Repeat, no drop."

He eased the plane into a broad left turn and prepared to come at the target again. This time, as the *Bockscar* approached Kokura, flak began to burst around them. In World War II, bombers rarely made a second run on their target, since that gave defenders time to zero in on an airplane. Sweeney took the plane up to 31,000 feet to try to avoid the flak.

"I can't see it," Beahan yelled again as they approached.

"No drop," Sweeney said. He had to follow orders.

As Sweeney wheeled the *Bockscar* around for a third pass, his radar operator, Abe Spitzer, said, "Zeros coming up. Looks like about ten." Sweeney flew the plane a thousand feet higher. He would try another angle—maybe the target would be visible from that direction. The flak was so close that it was causing the airplane to shudder. But it was hopeless—the clouds were impenetrable.

Sweeney had to decide. The flak was too close, the Zeros were approaching. After spending so much time waiting for Hopkins and making three runs on Kokura, the *Bockscar* was dangerously low on fuel. According to flight engineer John Kuharek, they no longer had enough gas to get back to Tinian. At this point, they would be lucky to get to the nearest US base on Okinawa. Spitzer, from his station in front of the radar equipment, asked, "What's wrong with Nagasaki?"

Sweeney had been thinking the same thing. He knew he couldn't drop the bomb visually on Kokura. His only option for a visual drop was the backup target. That's what he would do. He swung his plane south, away from Kokura—in the process almost running into Bock, whom he had not warned about the maneuver. Since that day, people in Japan have spoken of the "luck of Kokura"—being spared from annihilation by a threat that you did not even know was there.

Sweeney did not have enough fuel to fly to Nagasaki over water and avoid Japanese defenses. Luckily, the Zeros fell behind and no flak rose from below as he made his way across Kyushu. "Can any other goddamned thing go wrong?" he yelled to his copilot.

He turned to Kuharek. "What's the score?"

"Fifteen hundred gallons left, sir," the flight engineer responded.

"Jesus," said Sweeney, almost inaudibly.

A few minutes later, he told Spitzer, "Send Commander Ashworth up here."

When Ashworth arrived at the front of the plane, Sweeney said, "Here's the deal. We've got just enough fuel to make one pass over the target. Get that, one pass. If we don't drop it at Nagasaki, we may have to let it go in the drink. There's a slim chance that we might be able to make Okinawa with it still aboard, but the odds are damn slim."

Ashworth's face drained of blood. "What does that mean?"

"It means that, if you agree, we ought to do a radar run. . . . I'll guarantee we come within five hundred feet of the target."

Ashworth remained silent for a long time. Dropping the bomb by radar was strictly against orders. "Let me think it over, Chuck."

Ashworth walked back and forth in the crowded crew area behind the pilots, clasping and unclasping his hands. Who had the authority to make this decision was not clear. He could be overruled by Sweeney. But he was the weaponeer in charge of the bomb and had to make a decision, even if the decision was wrong. Below them, the forested mountains and settled lowlands of Kyushu slipped away to the north, bringing them ever closer to Nagasaki.

Finally Ashworth walked back to Sweeney. "We'll have to risk getting the bomb back to Okinawa," he said. "Our orders were that we weren't to make a radar drop. I'll have to carry out orders."

"Roger," said Sweeney. But Sweeney was deeply angry. He probably couldn't get the bomb back to Okinawa. More likely he would have to drop it in the ocean to have enough fuel to get to safety. But the wrath he would face from everyone on Tinian if he did that was unbearable.

The plane was five minutes from Nagasaki. In the crew area, Ashworth continued to agonize about what to do. If Nagasaki was cloudy, they were not supposed to drop the bomb there either. But what if they couldn't get the bomb back to Okinawa and it was lost, with no backup to replace it? What was more important—the orders he had received, or carrying out the mission?

He walked back to the front of the plane. "I've changed my mind, Chuck. We'll let it go over Nagasaki, visually, radar, or what have you."

"I could kiss you," Sweeney said.

They were two minutes away. Beahan began to line up the run on his bombsight. But as they approached Nagasaki from the southeast, they all could see that the weather, as at Kokura, had deteriorated. The central city, where they were supposed to drop the bomb, was obscured by clouds. Radar operator Jim Van Pelt had his face close to the five-inch radar screen. He was using it to locate the aiming point, in the middle of downtown Nagasaki, near the commercial district, surrounded by houses, with a row of elegant temples on the eastern edge of the city. He called out directions to Sweeney and Beahan so they would know when to drop the bomb.

Suddenly, from the front of the airplane, Beahan called out, "I've got it! There's a hole in the clouds. I can see the target."

Though downtown Nagasaki was covered with clouds, Beahan, in the very tip of the B-29's nosecone, could see that the valley northeast of downtown, where the Urakami River flowed, was partly clear. It wasn't the aiming point. But they were flying straight toward it, and Beahan knew from his previous study of potential targets that the valley was heavily populated and contained several Mitsubishi arms plants.

"You own it," said Sweeney.

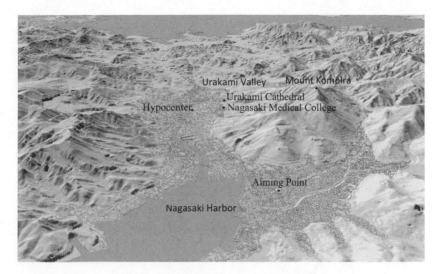

The atomic bomb dropped on Nagasaki exploded above the Urakami Valley, a section of the city northwest of the downtown commercial center. The Bockscar approached the city from the southeast. The distance from the original aiming point to the hypocenter—the point on the ground beneath where the bomb exploded—is about two miles. *Map courtesy of Matt Stevenson; drawing courtesy of Sarah Olson.*

Beahan called out the headings for Sweeney as he adjusted his Norden bombsight.

"Never again," Beahan muttered as he released the bomb. "Never, never again."

PART 3

UNDER THE MUSHROOM CLOUD

"This was the day of the apocalypse which Nagasaki will remember forever."

—Raisuke Shirabe

Chapter 16

NAGASAKI MEDICAL COLLEGE HOSPITAL

AT EXACTLY ELEVEN O'CLOCK IN THE MORNING ON AUGUST 9, 1945, Raisuke Shirabe was writing a research report in his office on the second floor of the surgery building at Nagasaki Medical College Hospital. He heard an airplane approaching. The air raid sirens had gone off at seven that morning, but the warning had expired at nine, and no further alarms had sounded. But this airplane was coming closer. It was obviously headed toward the Urakami Valley, not toward some other city in Japan. Shirabe rose from his desk and crossed the room to the door of his office. He hung his white lab coat on a hook and put on a dark suit jacket.

Just as he reached for the doorknob, the room filled with an unbelievably bright, bluish-white light. He immediately crouched near a sink next to the door. At that moment, the entire building shuddered. Parts of the wooden ceiling collapsed onto Shirabe's back, and a ferocious wind howled through the open windows. Everything in the room began to fall and swirl around him, as if he were in the middle of an immense windstorm. His eyes were tightly shut, but he could hear the roaring of the wind and feel objects falling on his back and head.

After a few moments he stood up and looked around. His office was completely dark. He heard a sound like that of heavy raindrops falling, which he thought might be dirt that had been lifted into the

Raisuke Shirabe, a surgeon at Nagasaki Medical College Hospital. *Courtesy of Nagasaki Association for Hibakushas' Medical Care.*

sky by the explosion. "I cannot describe my thoughts during this period," he later wrote. "It was like I had been left alone in the middle of hell."

Slowly the room began to lighten. Everything in his office—the bookcases, desk, bed, screen—had been knocked over. The contents of the room were scattered everywhere and covered with parts of the ceiling. Shirabe went to his desk. The paper he had been writing, his watch, his briefcase—all were gone, and he knew he would never be able to find them. Picking his way through the debris, he entered the hallway and walked down the nearby stairs. At the entrance to the building, he came across a woman whose appendix he had removed a couple of days earlier. She was standing with the support of a man but seemed otherwise unharmed. He briefly examined the woman and told her, "You're all right, don't worry."

The two dozen or so heavy concrete buildings that made up Nagasaki Medical College Hospital were on the lower parts of the slope leading toward Mount Kompira, on the eastern side of the Urakami

Valley. It was the oldest Western-style medical college in Japan, founded in 1857 by the Dutch physician Johannes L. C. Pompe van Meerdervoort, and had produced generations of doctors and nurses who had practiced throughout Japan. Even in the summer of 1945 it was thronged with students, though most would be drafted as soon as they graduated and sent to the war.

Shirabe walked quickly toward a bomb shelter in the hill behind the hospital, fearing that another bomb might drop at any moment. He began to come across bodies lying on the ground. He looked back at the hospital buildings. All the windowpanes had been blown out, and bodies were hanging from several of the open windows. How could a bomb have done all this?

Air raid shelters in Japan typically consisted of two caves dug into a hillside with a tunnel between them to accommodate more evacuees. The shelter behind Nagasaki Medical College Hospital was already filled with people. Inside, Shirabe met a nurse from his department who had a severe wound on her forearm. He took out his handkerchief and tied it around her elbow to stop the bleeding.

He decided that he needed to help others escape from the hospital, so he left the shelter and made his way toward the main building. But so many people were streaming out the door that he could not make his way inside. Many were headed up the hill to the east of the hospital, trying to get away from the destruction. He noticed, for the first time, that he seemed to have no injuries, unlike most of the people emerging from the building.

He met another nurse that he knew and a former student. The student did not seem to be injured, though he was using a cane for support. The nurse's face was covered with blood, and she was wounded at the hip. Go up the hill, he told them; he would join them later. Now he needed to get to his surgery building and help any of the staff who might still be inside to get out.

Walking across the tennis court behind the hospital, he met Professor Hasegawa. The professor was staggering and had a laceration between his eyes, but other than that he looked fine. Next he saw

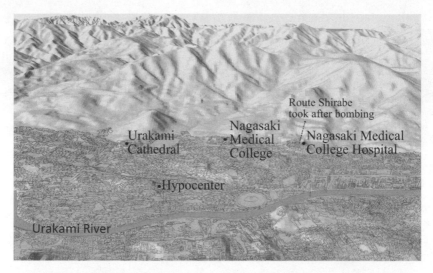

After the atomic bomb exploded above the Urakami Valley, Shirabe and other survivors from Nagasaki Medical College Hospital climbed the slope behind the hospital to escape the devastation. *Map courtesy of Matt Stevenson; drawing courtesy of Sarah Olson.*

Professor Ishizaki. He had severe burns on his face and upper body. The skin on his arms had come loose and was hanging in long, flapping shreds.

"Where were you?" Shirabe asked.

"I was in my office," Ishizaki responded.

"You'll be all right. Wait with Professor Hasegawa."

Near the surgery building he came across Dr. Kido and Head Nurse Murayama, who was also burned, though not as severely as Dr. Ishizaki. They told him that the nurses from his department were safe. Relieved, Shirabe turned to climb the hill with the others fleeing the hospital. A few minutes ago, this slope had been covered with the stems and leaves of sweet potato plants, which people had been growing to make up for the disastrous food shortages in Japan. Now the hillside was bare and brown. Nothing was left except the broken-off stems of the plants. The people making their way up the steep paths looked like tattered streamers against the muddy soil.

All around him, in a half-light of smoke and dust, were people who had been burned and injured. Many were naked, bleeding, blackened. They staggered, crawled, lay motionless on the ground. "Water," they called out. "Help me." "Mother." There was nothing Shirabe could do for them except call out encouragement. He kept climbing.

He heard his name being called. The president of the medical college, Susumu Tsuno'o, had been injured in the blast. Several faculty members and nurses had carried him up the hillside, and they were calling to Shirabe for help. When he joined them, he saw that the president's face had a bluish tint, and his shirt was covered with blood. "Where are your wounds?" asked Shirabe.

"My left arm and leg have small injuries," Tsuno'o answered in a weak voice. "I'll be okay."

Shirabe bandaged a wound on the president's leg with a piece of cloth. Tsuno'o looked uncomfortable in his bloodstained shirt, so Shirabe took off the president's shirt and replaced it with his own.

He turned to look at the hospital. Fires had broken out in the shattered buildings. They needed to get farther away. An instructor at the college lifted the president onto his back, and they resumed climbing. The president said, "I feel nauseated by brain ischemia when moved," and he vomited several times as he was being carried. By the time they got to a level spot by a boys' reformatory, which also was in flames, it was about one o'clock.

Then the wind changed direction and Shirabe could see beyond the hospital. It couldn't be. The entire valley, home to perhaps 100,000 people, was a sea of fire. Everything had been destroyed. Except for the reinforced concrete buildings of the hospital, designed to be resistant to earthquakes, not a structure seemed to be standing. No bomb could have done this.

Most of the people between the hospital and the center of the valley were already dead. Among those who were still living, scenes of unimaginable horror were taking place. One of the students at the medical college, having survived the blast, decided that he would make his way to another hospital several miles northeast of the medical school. Partway there he encountered an infant boy crying next to a demolished house. He heard the voice of a woman calling from under the fallen wood and tiles. "Help my child! Save him!" He could see part of the woman's legs but knew he could not pull her from the debris. "Don't worry," he shouted. "I'll take care of your child!" He picked up the infant and carried him away. Near the cathedral north of the medical school, where more than a hundred worshippers died when the ceiling and walls fell in, he and the infant child had to cross a small stream that traveled through a concrete-lined canal. It was choked with bodies. A woman and her newborn child, both dead, were still connected by an umbilical cord—she must have given birth when she entered the water. The medical student and child climbed up the far wall of the canal and continued walking. As they passed crumbled houses, they could hear the cries of pinned-down people burning to death.

On the hillside behind the hospital, Shirabe treated President Tsu-no'o's wounds with a first aid bag that someone had retrieved. The president's back was full of glass fragments, but Shirabe could do nothing about these. It began to rain. Shirabe thought about his son Koji, a student at the medical college. He would have been at a lecture when the bomb went off. Leaving the cluster of nurses, students, and doctors on the hillside, Shirabe made his way back down the slope. Near the medical school he again came across Dr. Ishizaki, who was lying on the ground and not moving. Someone had covered him with a quilt. Shirabe went to the burning lecture hall where his son would have been and began calling "Koji, Koji." But there was no reply. Other wounded students and faculty members were by the building. Some were near death, others barely injured. One helped him look for Koji, but they found nothing.

He heard a voice calling his name from the hillside. "Dr. Nagai's bleeding won't stop, please come." Takashi Nagai, later to become famous as the author of *The Bells of Nagasaki*, was lying on the hillside, with a deep cut on his right temple. Two assistants from the radiology department had been trying unsuccessfully to stop the bleeding. The only way Shirabe could stanch the wound was by pressing a sanitary napkin into the cut and suturing Nagai's skin over it. "I sensed a kind of godliness in the attitude of Dr. Nagai, who did not as much as frown during the operation without anesthesia," Shirabe later wrote.

Nagai, too, had witnessed horrific scenes that day. "Two children dragging the body of their dead father," he later recalled. "A young woman running frantically with the headless body of her baby in her arms; an elderly couple climbing the hillside hand in hand, panting; a young woman rolling, transformed into a fire ball when her scorched trousers were fanned into flames."

Through the smoke, the sun began to descend toward the hills on the other side of the valley. They would have to spend the night on the hillside. The survivors built fires and began to cook rice and radishes dug from the makeshift farms. Shirabe thought about the other members of his family. His oldest son, Seiichi, must have been working at the Ohashi ammunition factory in the northern part of the valley. Had the devastation extended that far? What about his wife, mother, and three daughters? The family had evacuated from the central part of the Urakami Valley a few months earlier to a rural area north of the city. Were they safe, or had they been killed, too?

The clouds above Nagasaki had cleared over the course of the day. As the sun sank in the west and the fires in the valley subsided, stars began to appear in the nighttime sky.

◉

GIVEN THE INTENT of destroying as much of Nagasaki as possible, Kermit Beahan could not have dropped the bomb containing Hanford's 13 pounds of plutonium in a much better place. It exploded 1,650 feet above the center of the Urakami Valley in a neighborhood of homes and small businesses halfway between two large Mitsubishi arms factories, both of which were destroyed. If he had dropped the bomb 15 seconds earlier or 15 seconds later, it would have gone off over the mountains on either side of the valley—still with devastating effects, but causing only part of the destructiveness that occurred.

In his office at the Nagasaki Medical College Hospital, Shirabe was 2,460 feet from the hypocenter, the point on the ground directly beneath the bomb's explosion. That means he was standing 2,960 feet—a little more than a half-mile—from the world's third nuclear explosion. From his office, the bomb exploded at an angle of 34 degrees above the horizon—about that of the mid-afternoon sun.

Shirabe was incredibly lucky. He later calculated that the radiation from the bomb went through two to three 12-inch-thick concrete walls to reach his office, which greatly reduced its intensity. Also, some of the gamma rays from a nuclear explosion arrive after the initial flash of light because they are generated by secondary reactions in the air surrounding the bomb. By crouching down next to the sink in his office when he saw the flash, Shirabe further reduced his body's exposure.

Other people in the hospital were not so lucky. They might have been near an open window, or on the side of the hospital closer to the bomb, or in the corner of a room where radiation could reach them through a wooden ceiling rather than a concrete wall. Anyone standing outside the hospital received a lethal dose of both gamma rays and neutrons. The exposure of those inside the hospital varied.

The bomb's shock wave destroyed virtually every building within a mile and a half of the hypocenter. The only buildings left standing after the blast were those of the hospital and several reinforced concrete schools likewise built to resist earthquakes. Wood from the demolished homes then provided fuel for fires that sprang up throughout the valley. Though Nagasaki did not experience a cataclysmic firestorm like the one that occurred in Hiroshima, much of the valley eventually burned, and fire even made its way into part of the downtown commercial core.

Along with its characteristic flash of visible light, an atomic bomb releases immense amounts of infrared radiation, which our bodies feel as heat. In Nagasaki, people near the hypocenter died instantly from the bomb's thermal radiation, and people as far as two-and-

Only reinforced concrete buildings, including those of Nagasaki Medical College Hospital, remained standing within a mile and a half of the hypocenter, which is in the valley beyond the hospital. Shirabe was in a building at approximately the center of these structures. *Getty Images.*

a-half miles from the hypocenter experienced serious burns. Dark colors in clothes absorb infrared radiation while light colors reflect it. Many people in Nagasaki had the colored patterns of their shirts, kimonos, and armbands burned into their skin.

People inside Nagasaki Medical College Hospital were partly protected from the bomb's radiation and blast. Still, the death toll there was horrendous. Of about 850 students, 650 died, partly because many were in the wooden buildings of the medical school rather than the concrete buildings of the hospital. Even inside the hospital, half the patients died, and almost half the professors. The people who streamed up the hillside after the bombing left many of their friends, colleagues, and family members behind.

BY TODAY'S STANDARDS, the atomic bomb dropped on Nagasaki was small. The most powerful weapons in the current US arsenal can release more than 50 times the energy of the bombs dropped on Japan. If a modern hydrogen bomb were exploded above the ridge that separates Nagasaki's two valleys, it would destroy the entire city, not just part of it.

Similarly, if a single hydrogen bomb were detonated above any city in the world, it would destroy much of that city. Houses within five miles of the hypocenter would be smashed, and people six-and-a-half miles away would suffer life-threatening burns. Moreover, even with recent reductions in the immense size of the US and Russian nuclear arsenals, multiple warheads target most American and Russian cities. After a large-scale exchange of nuclear weapons between the two superpowers, few people would emerge from those cities alive.

As with Hiroshima and Nagasaki, cities hit by nuclear weapons would then begin to burn. These massive conflagrations would inject large amounts of soot, smoke, and dust into the upper atmosphere, where the aerosols would block the light from the sun. The most recent studies indicate that a nuclear exchange of even 50 Nagasaki-size bombs would produce climate changes unprecedented in recorded human history and threaten the global food supply. A large-scale exchange of nuclear weapons would so reduce temperatures that most of the humans who survived the initial bombing would starve. Many people are concerned today that climate change poses a threat to human civilization, but the most certain and immediate threat still resides in the nuclear weapons now sitting in missile silos, bombers, and submarines around the world.

Chapter 17

THE URAKAMI VALLEY

WHEN SHIRABE WOKE THE NEXT MORNING ON THE HILLSIDE ABOVE THE
hospital, he saw that the fires had died down in the valley below
him, though tendrils of smoke still rose from the shattered buildings.
It was a clear and warm morning. On an adjacent part of the hill,
Nagai, who was a Christian, gathered with the other Christians on
the hillside by the small shrine where they had spent the night, faced
in the direction of the emperor's palace, and began to sing hymns.
"Their voices were well harmonized," Shirabe recalled. "I did not
go to the group's assembly line, but I sat up straight and joined the
group in my heart."

As the sun rose higher, Shirabe and a group of hospital staff began
carrying the injured, including the medical college president, back
down the hillside on stretchers. The president thought that patients
should be treated at the hospital, where medicines were available.
But when Shirabe ran ahead of the group to look for a suitable sick-
room, he found that all the buildings, including his own, containing
all his books and other possessions, had been gutted by fire. That
morning, injured people were still scattered around the former hospital, moving in a daze and calling for help. Shirabe wondered again
if his son might have survived the blast. He went to the auditorium
where he thought Koji would have been listening to a lecture, which

had burned to the ground, but he could tell from the size of the bodies he saw that none was his son.

He returned to the hospital and began treating patients. He bandaged wounds and injected people who were near death with heart-stimulating drugs. But their needs were endless and supplies soon ran out. He needed to get the injured away from the hospital and to a place where they could be treated properly. A few months earlier, he and his family had moved away from their home near the hospital to a small house in the northern part of the Urakami Valley. If it had not been damaged, he could treat patients there. He had some stores of medical supplies at the house. And he needed to see whether his family was alive.

Shortly after noon, he began to walk north through the ruined city. In the garden and pond of the house where his family had lived before the war, he saw charred corpses. The road was covered with shattered bricks and roof tiles still hot from the previous day's fire.

He began to pass into less damaged areas. The bomb had obliterated most of the Urakami Valley, but the damage was less severe to the north, especially where hills had shielded structures from the bomb's effects. His own house was on the north side of a hill, about two-and-a-half miles from the hypocenter. As he approached the house, he could see that it had incurred some damage. The glass windows had been blown out, and the sliding doors were broken. But the house was intact.

As he walked toward it, his wife Sumiko and all three of his daughters came running from the front door, tears in their eyes. He embraced them all. His oldest son, Seiichi, emerged last. He was alive but injured. He had been working in Mitsubishi's Ohashi ammunition factory in the northern part of the valley when the bomb exploded. His neck, back, and right forearm were severely burned where they had been exposed to the bomb's radiation through an open window, but that was the extent of his injuries. Doctors have a rule of thumb that if less than one third of the body's surface is

Shirabe with his mother Iso, his wife Sumiko, his three daughters Choko, Reiko, and Junko, and his sons Seiichi and Koji. *Courtesy of Hitomi Shirabe.*

burned, a patient should not die. Shirabe quickly did the calculation and concluded that his son would live.

Toward evening he went in search of a building that could serve as a relief station. A nearby school was too damaged to inhabit, but he found a building called the Iwaya Club that would work if the floor was repaired. He made arrangements with the building's manager to use it.

The next day, he walked back to the hospital and negotiated with two army officers, both graduates of the medical school, to use a truck to transport patients too injured to walk. He was able to find some medical supplies in the burned-out shell of the surgery building, and he used them to treat some of the injured. He also went to see President Tsuno'o, whose condition was worsening. All around the president, people who had not died in the initial blast were dying from their external injuries, from the rupture of their internal organs, and sometimes for seemingly no reason at all. Shirabe did what he could to help them. "I almost choked at the sight of continuous scenes of tragedy," he recalled.

There followed a stretch of days that passed in a blur. Shirabe was able to transport many students, nurses, professors, and other patients to the Iwaya Club, where he was able to care for them more effectively than at the hospital. He also was taking care of people in the neighborhood, using what supplies were available. But no matter what he did for them, strangely, many of his patients got worse rather than better. They had severe and bloody diarrhea, so much so that Shirabe thought they might have cholera and tried to isolate them. As his patients continued to sicken and die, the survivors devoutly cremated them in a nearby field.

On August 15, they heard that the emperor of Japan had ended the war. Shirabe was too busy to care. His son's condition had deteriorated over the previous days. A thick black substance oozed from Seiichi's wounds, and his face was darkening. Shirabe and his son both knew that he would soon die. Seiichi thanked his parents and apologized for his early death. He said that if he was reincarnated he would take revenge on their enemy. Though Shirabe knew that his son's condition was hopeless and that many other people needed his help, he stayed by his bedside. Seiichi died about noon the day after the emperor's announcement and was cremated on a nearby hill.

On August 22, President Tsuno'o died, surrounded by his wife and members of the hospital staff. His injuries had not been severe, but he had never recovered from the strange malady that had afflicted him and so many other people since the bombing. By this time, the people of Nagasaki realized that they were falling prey to a new and evil disease. They called it atomic bomb sickness. Right after the bombing, many survivors were nauseous, threw up, and had diarrhea. The fast-growing cells of their digestive tracts had been killed by the radiation, and their intestines no longer functioned. They became incredibly thirsty and lethargic, sometimes experiencing convulsions and delirium, and began to die. By the week of August

22, when President Tsuno'o died, people exposed to the bomb's radiation began to die from more widely distributed effects. Their mouths, noses, and rectums bled, and their skin was covered with blotches and spots where blood vessels had burst. They developed fevers and infections. There was little Shirabe and the other doctors who had survived the explosion could do for them.

On August 28, Shirabe, his wife, and his three daughters went to the hospital to find out what had happened to Koji, the second oldest son in the family. Shirabe guessed that he had been in the anatomy auditorium at the medical school, just north of the hospital. As they approached the burned-out school, they saw hundreds of crows flying in the sky. "Their voices were angry, as if they were cursing the souls of the dead," Shirabe remembered. Nothing was left of the auditorium except its concrete foundation. They saw mounds of human ashes in the foundation—students who had been listening to the lecture and were cremated in the bomb's fires.

Suddenly his daughter called out, "There's something here." She had found a remnant from a piece of clothing. They looked and saw the word "Yamamoto" written in black ink on the cloth. That was the son of Shirabe's eldest sister. He had been drafted to serve in the navy after completing medical school, and he had given his student uniform to Koji before he left. Finding the tag in the ashes meant that all hope was lost. Both of Shirabe's sons were gone.

By this time, Shirabe himself had fallen ill. He continued to trudge from house to house to treat patients in the neighborhood, but he had no energy and staggered when he walked. Finally even walking was impossible. Lying in bed, he found small purple spots on his upper arms and thighs. They were subcutaneous hemorrhagic spots from the radiation, he knew. He covered his arms and legs so that his wife would not see them. But when Sumiko glimpsed them anyway, she said, "It must be flea bites. I also have small ones." He wondered if he should make a will, but he was too weak to speak or even to turn over in bed. He thought that the end was near.

Then, after about a week, the spots began to change color—from purple to blue to yellow. By the second week in September, he was beginning to think that he might survive. Not long after that, a third-year medical student came to visit. He found a bottle of alcohol in a closet and asked, "Doctor, can I have a drink?"

Shirabe thought that it might be methyl rather than ethyl alcohol. "If you die as a result, don't hold me responsible," he said.

But the student drank some and did not become ill. "How about a drink, doctor?" he asked.

Shirabe was worried about damaging his radiation-ravaged liver, but he accepted a cup of alcohol from the student. It was delicious. He felt his body warm and his strength revive. Over the next few days, he drank a small amount of alcohol at breakfast and at dinner every day. That bottle of alcohol was "a savior," he later wrote. Sumiko noticed that he was looking better. By about September 20, he noticed that the spots on his arms and legs were gone.

Chapter 18

NAGASAKI

WHEN SHIRABE WAS A SCHOOLBOY GROWING UP IN A SMALL TOWN IN western Japan, he loved to learn. He used to stand outside the classrooms of older students so that he could hear their teachers lecture, sometimes calling out answers to the teachers' questions from the hallway. He never lost that curiosity. It carried him through medical school in Tokyo, through his first jobs as a professor and surgeon in Beijing and Seoul—where the higher salaries helped him support his growing family—and then into his appointment as first professor of surgery at Nagasaki Medical College Hospital.

After he recovered from atomic bomb disease, Shirabe realized that he now could learn from the people of Nagasaki. That fall he organized a team of other physicians and medical students, and together they developed a survey that they could use to determine the medical effects of the bomb on the residents of the Urakami Valley. Over the next few months, Shirabe and his colleagues interviewed more than 5,000 survivors, gathering information on their burns, injuries, and radiation sicknesses. Their results were among the earliest Japanese-generated data on the effects of atomic bombs on humans.

They found that within one kilometer of the hypocenter (nearly two-thirds of a mile), almost everyone died, even the few people without obvious injuries. The death rate was lower at the hospital because of its concrete walls, but at the nearby medical college, which consisted largely of wooden buildings, the mortality rate was almost 100 percent. After the deaths caused by the initial blast, the death rate rose to a peak on August 17 and then declined. These deaths "may represent death from secondary shock," Shirabe and his colleagues wrote in their initial report on the survey, before the effects of radiation on the human body were more fully understood. "This may be an interesting subject for study."

Those who were farther away from the hypocenter tended to live a week or two longer than those nearer the blast, but then their mortality rates rose as well. The time to death was greater in those younger than 16 and slightly greater in females than males. Some people as far as two-and-a-half miles from the hypocenter experienced symptoms of what would eventually be recognized as radiation sickness, with the rates declining in the few people surveyed who were farther away.

Among the survivors who were surveyed, external wounds were more common than burns, with burns somewhat more common in older people who survived than younger people. Almost all the burns among survivors were second-degree burns that left grotesque scars when they healed. The most common injuries among survivors were cuts from broken glass, followed by contusions where they were hit by objects or were propelled into objects.

The data Shirabe and his colleagues compiled were invaluable in understanding the effects of an atomic bomb on humans, but they did not become publicly available for years. When the US Army took control of Japan in September, it imposed strict censorship on the dissemination of all information from the country, and especially news regarding the atomic bombs. US scientists had largely failed to anticipate the effects of radiation sickness on the survivors of the initial blast, and the military did not want that information to become known. Groves dismissed the few stories that did emerge as propaganda designed to increase sympathy for the Japanese. "The atomic bomb is not an inhumane weapon," he told the *New York Times*. "I think our best answer to anyone who doubts this is that we did not start the war, and if they don't like the way we ended it, to remember who started it."

Censorship applied to everyone, not just to Japanese scientists. In early September, a reporter for the *Chicago Tribune* named George Weller managed to extricate himself from a press junket in southern Japan and make his way by train to Nagasaki. There he pretended that he was a US colonel and demanded food, lodging, and transportation. Every night he typed his stories by lamplight and sent them in a package labeled "Chief Censor, American Headquarters" to Tokyo, where he hoped MacArthur's censors would allow the stories to get out. "Today the writer spent nearly an hour in 15 deserted buildings of the Nagasaki Medical Institute hospital which sit on a hill on the eastern side of the valley," he wrote on September 8. "Nothing but rats live in the debris-choked halls." His hopes for publication were in vain. The censors suppressed all his stories and appear to have destroyed the copies they received. Weller kept carbon copies, but they were lost for 60 years until his son found them in a moldy box a few months after his father's death.

In the fall of 1945, the survivors of the Nagasaki bombing began trying to create new lives for themselves. Almost all had lost family members, and many were injured or sick. Many men working in munitions factories in the Urakami Valley were killed, leaving their wives and children destitute. Children who had lost both parents lived on the streets, begging, stealing, and going through garbage bins for food. Without income or any other means of support, many people treated their injured family members at home. Funeral pyres burned for months as bulldozers began to scrape the valley clean.

The postwar years were a time of great misery for Nagasaki. As writer Susan Southard recounts in her book *Nagasaki: Life After Nuclear War*:

Fourteen or fifteen people often lived in a single room with no furnishings. Running water was still not available, so survivors hauled springwater from the mountains and collected rainwater to boil and drink. Without toilets, people dug holes in the ground outside their shanties and covered them with wooden boards. Without bathtubs, they heated water in large oil drums and bathed standing up. To battle the winter winds, families wore as many layers of donated clothing and blankets as they could, huddling beneath umbrellas around wood-burning hiba-chi to protect themselves against the rain, sleet, and snow that fell through their makeshift roofs.

The psychological damage was as grave as the physical damage. Many survivors remembered watching those around them die and being unable to help. Many had disfiguring scars on their faces or bodies. Those who survived without visible injuries or scars were terrified that they had been damaged by the bomb's radiation and would die of the diseases that had killed so many people they knew. They thought that any children they might have would be deformed, and many refused to marry or were rejected as mates. Those who survived the atomic bombings came to be known as *hibakusha*, which means explosion-affected people. They were shunned by other people in Japan, and many moved to other parts of the coun-try and tried to hide what had happened to them.

In 1949 the Japanese legislature passed the Nagasaki Interna-tional Culture City Construction Law, which provided funding to help rebuild parts of the city that had been damaged. Censorship began to ease. Also in 1949, three years after it was written, Takashi Nagai's memoir *The Bells of Nagasaki* was published and became an international bestseller. In the book, Nagai recounts the bombing and how it killed his wife and many of his colleagues, students, and patients. He details his own injuries and radiation sickness and the efforts he made to help other survivors. He describes several young men digging up the bells of the ruined cathedral and ringing them as a message of peace and understanding. "Men and women of the world, never again plan war," Nagai wrote at the end of his book. "With this atomic bomb, war can only mean suicide for the human race. From this atomic waste the people of Nagasaki confront the world and cry out: No more war!"

Every year since the bombing the city has held commemorative events on August 9. In 1949, for the first time since the war, thousands of Catholics from Nagasaki and elsewhere marked the anniversary in the ruins of Urakami Cathedral. In the 1950s, the first new buildings opened on the grounds of what was now the hospital of the Nagasaki University School of Medicine. Since then, the remains of the old hospital have been replaced by much larger and more modern buildings.

SHIRABE SPENT THE REST of his long and productive career at the Nagasaki University School of Medicine. In the 1950s, after becoming director of the Nagasaki University Hospital, he established a relationship with the Atomic Bomb Casualty Commission, which later became the Radiation Effects Research Foundation. The commission was established in 1946, at President Truman's request, by the National Academy of Sciences and its operating arm, the National Research Council, to study the effects of the atomic bombs on Japanese survivors. Though it was widely criticized in Japan and elsewhere for not treating the injuries and illnesses of the *hibakusha*, the commission provided a way for Shirabe's early studies and other medical investigations to become public. Today a small museum in the medical college still proudly displays the results of the work that Shirabe and other college physicians and students did after the war.

Shirabe, like other people in Nagasaki, also committed himself to remembering what happened on August 9, 1945. He became the head of an association of families of students killed by the bomb, and he edited a collection of writings on the experiences of *hibakusha* that eventually ran to seven volumes. He served for eight years as visiting director of the Radiation Effects Research Foundation while continuing to study the long-term health effects of the bomb's radiation. He died in 1989, a month before his ninetieth birthday, and soon thereafter was named an honorary citizen of Nagasaki.

Today, the Urakami Valley has been completely rebuilt, and

the city at large contains relatively few reminders of the devastation caused by the atomic bomb. A gray stone monolith in a small
park surrounded by trees marks the point beneath where the bomb
exploded. A larger park on an adjacent hill, on the site of a prison
that was destroyed by the bomb, contains sculptures and monuments
sent by governments around the world to celebrate and preserve
peace. The Atomic Bomb Museum halfway between the Peace Park
and the rebuilt medical college displays artifacts recovered from the
wreckage and explains what the atomic bomb did to the city. To
the north of the medical college, next to the rebuilt Urakami Cathedral, one of the original cathedral's bell towers is still lodged into
the earth on the banks of the concrete canal that was choked with
bodies on August 9, 1945. The rest of the city hums with homes,
businesses, traffic, and tourists.

Yet even today, in quiet corners of the city, vivid reminders of that
long-ago event remain. In the *tatami* room of a house just a mile or
so north of the medical school, where Shirabe's now elderly daughter
Choko still graciously greets inquisitive visitors, two oil paintings
of young men in uniform hang on the wall. They are her brothers,
Seiichi and Koji, killed by the atomic bomb dropped on Nagasaki.

PART 4

CONFRONTING ARMAGEDDON

"The total elimination of nuclear weapons is the only absolute guarantee for all of us to be safe from the threat of nuclear annihilation."

—Jim Stoffels, peace activist

Chapter 19

THE COLD WAR

AS IN THE REST OF THE UNITED STATES, THE IMMEDIATE REACTION IN Richland to the news that an atomic bomb had been dropped on Japan was largely elation. "It's Atomic Bombs," proclaimed a special edition of *The Richland Villager* on August 6, 1945. "It is the greatest force ever harnessed—and may change the course of civilization."

For people across the country, the moment they heard about the atomic bomb was seared into their memories, just as later historic events would be. President Truman announced the bombing of Hiroshima in a radio address at 11:00 a.m. Washington, DC, time, or 8:00 a.m. in Richland. By then, most of the men had already left for the site, so their wives were the first to hear. They called each other "in a flurry of phone calls which kept the switch boards humming," *The Villager* reported. A few were able to reach their husbands at work, and word began to spread through the fuel fabrication facilities, the reactor control rooms, the separation plants. That morning, before leaving for work, Matthias had told his wife to listen to the radio. She was grinding ham for dinner when she heard the news. She burst into tears. "I couldn't help but believe that God, wearying of this long and tortuous war, had finally, reluctantly, given us this terrible weapon with which to end it."

The reporting was murky at first—even Matthias was initially unclear about whether the bomb dropped on Hiroshima had used uranium from Oak Ridge or plutonium from Hanford. But the story quickly came into focus. An entire Japanese city of a quarter million people had been destroyed. Tens of thousands, possibly more than 100,000, were dead.

People were shocked, excited, hopeful that the war would soon be over—and frightened. An atomic bomb had destroyed a Japanese city, which meant that atomic bombs could destroy other cities, including American cities. Even Truman's initial statement hinted at that possibility: "Under present circumstances it is not intended to divulge the technical processes of production or all the military applications, pending further examination of possible methods of protecting us and the rest of the world from the danger of sudden destruction." On the evening of the Hiroshima bombing, NBC news commentator H. V. Kaltenborn said in his radio broadcast, "For all we know, we have created a Frankenstein! We must assume that with the passage of only a little time, an improved form of the new weapon we use today can be turned against us." Two days after the Hiroshima bombing, the *Milwaukee Journal* published a map showing how much of Milwaukee would be destroyed by an atomic bomb. Radio newscaster Don Goddard observed that the bombing of Hiroshima was like "Denver, Colorado, with a population of 350,000 persons, being there one moment, and wiped out the next."

The bombing of Nagasaki just three days after the destruction of Hiroshima created another surge of excitement and fear. The United States seemed to have a ready supply of atomic bombs. Certainly the Japanese could not hold out for long under such an onslaught. In fact, the plutonium and high explosives for a third bomb were still in Los Alamos, though Groves was getting ready to send them to Tinian. But the day after the Urakami Valley was destroyed, President Truman ordered a halt to any more atomic bombings. As Henry Wallace, Truman's Secretary of Commerce, wrote in his diary, "Truman said he had given orders to stop atomic bombing. He said the thought of wiping out another 100,000 people was too horrible. He didn't like the idea of killing, as he said, 'all those kids.' "

On August 15, *The Richland Villager* released another two-page special edition with the headline filling half the front page: "Peace!

Our Bomb Clinched It!" The message of surrender Emperor Hirohito had broadcast to the Japanese people just a few hours earlier
seemed to confirm the newspaper's conclusion:

> The enemy has begun to employ a new and most cruel bomb, the
> power of which to do damage is, indeed, incalculable, taking the
> toll of many innocent lives. Should we continue to fight, not only
> would it result in an ultimate collapse and obliteration of the
> Japanese nation, but also it would lead to the total extinction of
> human civilization.

The vast majority of Americans supported dropping atomic bombs
on Japan. When a sample of Americans were asked later in the fall
which of four statements best captured their opinion about the
bombings, the majority agreed that the United States should have
used atomic bombs on two cities, just as it did. Nearly a quarter
were even more aggressive, agreeing with the statement that "we
should have quickly used many more of the bombs before Japan had
a chance to surrender." Less than 20 percent thought that atomic
bombs should not have been used or should have been demonstrated
on an unpopulated region. Fear and hatred of the Japanese, stoked by
news stories and propaganda since the attack on Pearl Harbor, were
widespread in the United States. Newspaper stories, radio commen-
taries, and newsreels had documented the atrocities committed by
Japanese soldiers across Asia, including the torture and mutilation
of prisoners and the rapes and killings of civilians. Almost 100,000
Americans died in combat in the Pacific theater during World War
II, and many families feared for their loved ones if US troops would
have to invade mainland Japan to end the war. Emperor Hirohito's
announcement that Japan had surrendered was greeted with widespread celebration and relief.

   Yet from the earliest days, some people began to ask a question
that has been debated ever since: Were the bombings of Hiroshima
and Nagasaki necessary to force Japan's surrender? The fact that

the surrender came just six days after the Nagasaki bombing seems to clinch the case. But many other things were going on toward the end of the war. Because of the naval blockade and aerial bombardment of industry and railroads, the Japanese economy was on the verge of collapse, with severe shortages of food, fuel, and industrial materials. Truman and his aides knew that top Japanese officials were searching for ways to surrender, though the empire's military leaders were vowing to wage a final battle against an invading army. Many US military commanders and soldiers thought that an invasion of Japan would be necessary, but they did not know how close Japan was to defeat or that atomic bombs existed. The entry of the Soviet Union into the war on August 8 forced the Japanese leaders to recognize that their position was hopeless. The advent of atomic bombs gave them an excuse to surrender, even though they initially made little distinction between the use of atomic or incendiary weapons. Groves's single-minded determination to finish and use the bombs, combined with scientists' fascination with the technology, gave the project a momentum that was hard to reverse. Meanwhile, Truman and his future secretary of state, Jimmy Byrnes, were eager not only to end the war as soon as possible but also to demonstrate the power of the atomic bomb to the Soviet Union, which they clearly saw as a future adversary.

The scholarship on what one historian has called "the most controversial issue in all of American history" is vast, and the question of whether the bombings could have been avoided will never be answered definitively. The consensus view of most though not all historians who have studied the issue is that an invasion of Japan probably would not have been necessary to end the war, even if atomic bombs were not ready to drop on Japan in August 1945. But most Americans believe something quite different. Partly because of a postwar public relations campaign carried out by the public officials involved in the decision, Americans largely think that Truman faced a stark choice between atomic bombs and an invasion. That is not true, and it obscures the many factors in play in the summer of

1945. At the same time, a path by which the US government would *not* have used atomic bombs when they became available is difficult to envision. Whether dropping atomic bombs on Hiroshima and Nagasaki was necessary will always remain one of the great "what ifs" of human history.

Yet the decision left one undeniable legacy. The idea that the atomic bombs were necessary to end World War II helped create a conviction that nuclear weapons could serve a useful purpose in warfare. They could win a war that threatened American lives and American interests. Granted, that victory could come only at the price of enormous destruction and moral anguish. But if all else failed, nuclear weapons were the ultimate fallback. The United States has never renounced the first use of nuclear weapons, even if their use risks the end of human civilization. Would that be the US stance if Americans did not so forcefully believe that the bombings of Hiroshima and Nagasaki were necessary to end World War II?

THE MAN WHO DISCOVERED plutonium less than five years before it fueled the Nagasaki bomb was largely excluded from the official deliberations on its use. Glenn Seaborg had been a member of the Franck committee and also had written a letter to Ernest Lawrence arguing "not to use the weapon on Japan without warning but to demonstrate the weapon in the presence of all leading countries including Japan." However, he never renounced the subsequent bombing of Japan. "I understand and do not quarrel with Truman's decision to use the bomb on Hiroshima (though I'm not convinced that the follow-up bombing of Nagasaki was necessary)," he later wrote. He agreed with the Scientific Advisory Panel that the Japanese might have interfered with a demonstration or that the failure of a bomb might have emboldened the Japanese. And, as with many other Manhattan Project scientists, he had personal connections to the war. "Some of my cousins whom I'd grown up with in South Gate were stationed in the Pacific islands, preparing for the invasion of Japan, which they were certain would be necessary to force Japan's surrender. . . . For years after the war, at family reunions they made a point of thanking me for my work—they were convinced that the bomb saved their lives. That gave our achievement personal meaning."

Among the other scientists who designed and built the atomic bombs, reactions to the bombings were varied. Three hours after President Truman's announcement about Hiroshima, Groves called Oppenheimer in Los Alamos. "I'm proud of you and all your people," he said.

"It went all right?" Oppenheimer asked.

"Apparently it went with a tremendous bang."

"Everybody is feeling reasonably good about it," Oppenheimer said. "I extend my heartiest congratulations. It's been a long road."

That evening, Oppenheimer spoke in a Los Alamos auditorium before a cheering crowd of scientists, soldiers, and support staff. At one point, according to a young physicist who was there, he clasped his hands, raised them over his head, and shook them like a prizefighter. He said that he was proud of what they had accomplished together. His only regret was that they had not built atomic bombs in time to use them against the Germans.

But the mood at Los Alamos quickly darkened. When reports from Hiroshima and Nagasaki began to filter back to the plateau, Oppenheimer and many of his colleagues became increasingly depressed. "As the days passed," the historian Alice Kimball Smith later wrote, "the revulsion grew, bringing with it—even for those who believed that the end of the war justified the bombing—an intensely personal experience of the reality of evil."

The news of the Nagasaki bombing just three days after Hiroshima, before the Japanese leaders had a chance to fully assess the damage and their options, was especially troubling. Like Seaborg, other Manhattan Project scientists expressed doubts about the second bombing, even if they supported the first. Physicist Samuel Allison, who had called out the countdown for the Trinity test, said a few weeks later, "When the second bomb was released, we felt it was a great tragedy."

All the members of the Scientific Advisory Panel, not just Oppenheimer, struggled with their feelings. Outwardly, Arthur Compton, the chair of the National Academy of Sciences committee that had recommended going forward with the project, did not betray ambivalence about the bombings, despite his religious convictions. "Before God our consciences are clear," he later wrote to a friend. "We made the best choice for man's future that we knew how to make." Yet he obviously dwelled on the issue after the war and addressed it at length in his 1956 book *Atomic Quest: A Personal Narrative.* In a section entitled "Choice," he reiterated the various reasons for dropping atomic bombs cited by political leaders after the war—use of the bombs could prevent an invasion of the Japanese homeland, they would make it possible for the Japanese to surrender with honor, the bombing of Nagasaki sent a message that the United States would continue to use atomic bombs if necessary. But the confusion of a critical paragraph in an otherwise cogently argued book may hint at an underlying ambivalence: "We recognized that victory was necessary if the danger of subsequent Japanese aggression was to be eliminated. The one and only compelling motive for our use of the bomb was to bring a stop to the killing needed to achieve such victory."

Ernest Lawrence, whose discovery of the cyclotron had greatly accelerated the pace of nuclear research in the 1930s, found solace in the idea that nuclear weapons would make war obsolete. "As regards criticism of physicists and scientists, I think that is a cross we will have to bear," he wrote a friend. "In the long run the good sense of everyone the world over will realize that in this instance, as in all scientific pursuits, the world is better as a result." Yet he also feared an arms race and believed that the United States must stay ahead of all other nations by quickly developing and stockpiling atomic weapons—an activity he pursued intensely after the war.

Enrico Fermi, whose discoveries contributed more to the Manhattan Project than those of any other scientist, said hardly anything after the war about the bombings. He had learned in Italy not to comment publicly on political matters. But he still had strong ties to the Chicago scientists and heard from them often. He also heard from people outside the United States who had a different perspective than most Americans. From Italy his sister wrote of the bomb: "All [here] are perplexed and bewildered by its dreadful effects, and with time the bewilderment increases. . . . For my part I recommend you to God, Who alone can judge you morally." Not until years later did Fermi reveal a bit of his thinking about the development of ever more powerful weapons, and even then with his usual whimsy: "I expect to sleep as well as my insomnia permits. I am a fatalist by nature anyway."

As might be expected, Groves never publicly expressed any hesitancy or regret about how the atomic bombs he had built were used. Shortly after the war, he said, "I have no qualms of conscience about the making or using of it. It has been responsible for saving perhaps thousands of lives. If the bomb had not been used the Japs would have held out for 60 to 90 days longer. We knew what that would mean in the sacrifice of human lives." Only one slight hint has ever emerged that Groves might have questioned the bombing of Nagasaki—and even then his concerns seem mostly tactical rather than humanitarian. Many years later, Alvin Weinberg, a physicist who helped design the Hanford reactors, told an interviewer: "I don't think the second bomb, the Nagasaki bomb, was necessary. No, that was strictly General Groves. In fact, I was drinking whiskey with Groves one night in a hotel bar in New York after the war and he more or less agreed the second bomb was not needed. He didn't say why. Groves was first of all a military man."

Nor did Truman ever publicly express any regrets about his role in dropping the atomic bombs, going so far as to say that he "never lost any sleep over that decision." But Truman was also a highly religious man, and it seems unlikely that he closed his heart entirely to the people of Hiroshima and Nagasaki. For the rest of his life, he wildly exaggerated the number of US soldiers his military officers thought could die in an invasion of Japan, citing figures as high as half a million American lives. After his death, a researcher discovered that Truman had collected many of the books written about atomic bombs in his library. In one of them, he had underlined portions of Horatio's soliloquy from *Hamlet*:

> *Let me speak to th' yet-unknowing world*
> *How these things came about. So shall you hear*
> *Of carnal, bloody, and unnatural acts,*
> *Of accidental judgments, casual slaughters*
> *Of deaths put on by cunning and forced cause.*
> *And, in this upshot, purposes mistook*
> *Fall'n on the inventors' heads.*

THE CHEMISTS, PHYSICISTS, METALLURGISTS, and others at the Met Lab in Chicago heard about the bombings the same way other people did—on the radio, through breathless phone calls, in agitated workplace conversations. But they knew better than most what the development of atomic bombs meant for the United States. "In the summer of 1945," wrote Eugene Rabinowitch, who had been on the Franck committee with Seaborg and Szilard, "some of us walked the streets of Chicago vividly imagining the sky suddenly lit by a giant fireball, the steel skeletons of skyscrapers bending into grotesque shapes and their masonry raining into the streets below, until a great cloud of dust rose and settled over the crumbling city."

The man who had first conceived of how an atomic bomb might work wrote a friend that he considered their use against Japan "one of the greatest blunders of history." Now Szilard set about trying to repair the blunder. He and his colleagues were hampered in what they could say and write because of the secrecy provisions imposed by Groves, who even classified Szilard's petition to Truman secret when Szilard sought to release it in *Science* magazine. But gradually they began to find sympathetic editors, journalists, and congressmen they could lobby. He and his colleagues in Chicago and elsewhere were determined to do what they could to shape the postwar uses of the bomb, even if they had failed during the war.

They based their activism on three premises evident in their earlier writing. First, other nations would soon develop atomic weapons. The Manhattan Project scientists knew their European counterparts from before the war. They understood that the Soviet Union had scientists and engineers who were just as good as those in America. Russian scientists and engineers would not take much longer to construct a bomb than the Manhattan Project had.

Second, no defense against atomic weapons was possible. They were sufficiently small to be smuggled into any nation, and a single bomb could destroy a city. Furthermore, the Germans had demonstrated, with their V-2 rockets, that explosives could be lofted from one nation to another. The combination of ballistic missiles and nuclear weapons would be unstoppable. Perhaps a defensive system would someday be able to intercept some of the missiles, but the rest would get through.

Third, the only way to protect nations from the terror of nuclear weapons was through some form of international cooperation. Perhaps an international agency under the United Nations, created earlier that year in San Francisco, could control the materials needed to make atomic bombs. Or maybe the United States and other countries could open their laboratories and factories to international inspectors who could ensure that bombs were not being made. It was even conceivable that some sort of world government could take over some of the governmental functions previously conducted by national governments, including anything connected to atomic bombs or nuclear energy. Regardless of the approach taken, some sort of international collaboration and control was essential, the Chicago scientists believed. Otherwise, the outcome was obvious. The nations of the world would begin building and stockpiling nuclear weapons for use against their enemies. And as the number of nuclear weapons in the world grew, some of them eventually would be used.

Activist scientists needed an organization to marshal their efforts. On September 25, what would soon be known as the Atomic Scientists of Chicago—which later expanded into the Federation of Atomic Scientists and then the Federation of American Scientists—elected a seven-member executive committee that included Szilard, Seaborg, and two other members of the Franck committee. Oak Ridge and Los Alamos soon were in the process of forming similar groups, but the Chicago scientists took the lead.

They didn't have long to encounter their first major challenge. Now that the war was over, legislation would be needed to govern the control of atomic energy in the United States. At the beginning of October, Senator Edwin Johnson of Colorado and Representative Andrew May of Kentucky introduced a bill that would put atomic energy under the control of a nine-member commission. Drafted largely by two lawyers in the War Department with input from Bush, Conant, and others, the bill gave the commission sweeping powers to establish and operate facilities, regulate all forms of nuclear research, and punish those who violated its directives—responsibilities that Groves and the Manhattan Engineer District had had during the war.

The more the Chicago scientists read of the draft bill, the more alarmed they became. The May-Johnson bill called for commission members to be part-time and appointed by the president. Members of the military could serve on the commission or as its administrator or deputy-administrator, positions that would have great power if the commission members served just part-time. The commission could label any information it wanted secret, with severe penalties, including up to 30 years in jail, for violations of security regulations. The bill did not address the international control or sharing of nuclear information, only restrictions on how that information could be used and disseminated.

To the Chicago scientists, the intent of the bill was obvious. This was an attempt by Groves and his allies in government to run America's postwar nuclear program the same way they had run the Manhattan Project—with strict secrecy, compartmentalization, little input from scientists or the public, and as little international cooperation as possible. America's political and military leaders obviously thought that the United States could monopolize knowledge about atomic weapons, or at least stay comfortably ahead of other nations. The bill was a sure recipe for an international arms race.

The Chicago scientists and their allies mobilized to defeat the bill and advance the cause of international control, but what approach would be most effective? They had several choices, such as emphasizing the advantages of a strict international regime governing nuclear weapons. Instead, they decided to stoke people's fears.

The tactic seemed obvious at the time. If anything, public alarm and despondency deepened in the months after the war. One anthropologist described America as being in the grip of a "fear psychosis." An article in *Life* magazine published in November 1945 and entitled the "36-Hour War" was emblematic of the mood. The article opened with a spread of a nuclear bomb detonating above Washington, DC. It then described a future nuclear war in which intercontinental ballistic missiles rained down on American cities. Antiballistic missiles fired from the United States stopped a few of the incoming missiles, but most got through. The magazine observed that 40 million Americans would be killed and most of the nation's cities leveled. The final page showed technicians testing the rubble of the New York Public Library for radioactivity, with only the proud lions at the former library's entrance left standing.

Scientists arguing against the May-Johnson bill sought to intensify people's anxiety. A March 1946 book published by the Federation of Atomic Scientists entitled *One World or None* included essays by Robert Oppenheimer, Albert Einstein, Niels Bohr, and other scientists on the threat posed by atomic weapons. The first essay was by physicist Philip Morrison, who had helped load the bombs onto the *Enola Gay* and *Bockscar* and later went to Hiroshima to survey the

An illustration in the November 19, 1945, issue of *Life* magazine depicted the effects on New York City of a nuclear attack on the United States. *Courtesy of Noel Sickles.*

damage. Entitled "If the Bomb Gets Out of Hand," Morrison's essay described what would happen if an atomic bomb were detonated a half-mile above Third Avenue and East 20th Street in New York City. "From the river west to Seventh Avenue, and from south of Union Square to the middle Thirties, the streets were filled with the dead and dying," he wrote. Nor did those farther away escape harm:

The most tragic of all the stories of the disaster is that of the radiation casualties. They included people from as far away as the Public Library or the neighborhood of Police Headquarters downtown, but most of them came from the streets between the river and Fifth Avenue, from Tenth or Twelfth to the early Thirties. They were all lucky people. Most of them had had remarkable escapes from fire, from flash burns, from falling buildings. . . . But they all died. They died in the hospitals of Philadelphia, Pittsburgh, Rochester, and St. Louis in the three weeks following the bombing. They died of unstoppable internal hemorrhages, of wildfire infections, of slow oozing of the blood into the flesh. Nothing seemed to help them much, and the end was neither slow nor very fast, but sure.

Altogether, predicted Morrison, 300,000 New Yorkers would be killed by a Nagasaki-sized bomb, and about an equal number seriously injured. Furthermore, this was the consequence of just a single bomb. As Morrison warned:

> The bombs will never again, as in Japan, come in ones or twos. They will come in hundreds, even in thousands. Even if, by means as yet unknown, we are able to stop as many as 90 percent of these missiles, their number will still be large. If the bomb gets out of hand, if we do not learn to live together so that science will be our help and not our hurt, there is only one sure future. The cities of men on earth will perish.

●

AMIDST THIS POLITICAL MANEUVERING, the residents of Richland awaited the decisions that would determine their futures. At the end of World War II, about 15,000 people lived in Richland, with about the same number living in Kennewick and Pasco a few miles farther down the Columbia. Almost all of them were there because the federal government had chosen to build a factory in the desert to produce plutonium. Now they wondered what would happen next. Would the government shut down Hanford? If so, the towns would likely shrivel up and blow away, as so many western towns had done in the past.

Right after the war, Groves had ordered production at Hanford scaled back. But the three reactors continued to operate, and the T and U Plants continued to separate plutonium from the irradiated slugs. By the end of the year, Groves expected to have 20 Nagasaki-type bombs in the US nuclear arsenal.

News reports after the war had discussed both plutonium and implosion, but Groves and Matthias continued to insist that Hanford's workers maintain absolute secrecy. "Discussion of process, production or the employment of the Atomic Bomb should be limited," Groves proclaimed. Hanford employees had to sign a statement saying they would not talk about what they saw, heard, or did at the site, even with family members. Military intelligence continued to open mail, listen to phone calls, and otherwise monitor the lives of Hanford workers. Local police had copies of the key for every house in Richland, photographs required the approval of the area manager, and even the phone book was classified. Right after Hiroshima and Nagasaki, the town was filled with journalists who wrote stories about Richland as the "atomic village" and a "model residential city." "Richland is News Center of the World: Army Lifts Curtain on Village Plant," *The Richland Villager* enthused, but the journalists left after a few days. By the end of 1945, the people of Richland were hard at work and worried about their jobs, just like the people in any factory town. They just happened to make plutonium.

IN THE FALL OF 1945, the Chicago scientists and their allies led a massive lobbying campaign against the May-Johnson bill. They criticized the bill's many flaws in congressional hearings. They gave interviews to journalists about what would happen if the bill passed. They convinced other scientists and engineers and other organizations to oppose the bill, thereby widening their movement. In December, they began publishing the *Bulletin of the Atomic Scientists*, which remains a powerful voice on nuclear issues today.

The public was impressed by the sincere young physicists who had won the war. By the end of 1945, public sentiment had turned sharply against the bill. Even President Truman, initially a backer of the May-Johnson bill, had withdrawn his support.

On December 20, 1945, with the bill all but dead, the junior senator from Connecticut, Brien McMahon, introduced a new proposal to control the future of atomic energy in the United States. Strongly influenced by the atomic scientists, this bill was much more to their liking. The security provisions were less onerous; scientists did not have to fear that their laboratories would again be turned into armed camps. The commission members would be full-time and appointed by the president, which in theory would reduce military control of the commission. The commission also could work toward international control of atomic energy, though the proposed legislation did not specify how this was to be done.

The McMahon bill faced powerful opponents, and none was more visible than Leslie Groves. Groves knew that the military could not continue to control nuclear weapons in the way he had controlled them during the war. Yet he was convinced that the military needed to have a dominant role in decisions about nuclear weapons, even if authority ultimately resided in the president. Groves was convinced that the Soviet Union and other countries would take many years to build atomic bombs. He knew how much effort, creativity, and industrial capacity it had taken for the Manhattan Project to succeed, and he did not think the Soviets capable of a comparable achievement. If the proper legislation could be passed, the United States could maintain a monopoly over nuclear weapons well into the future and leverage its nuclear strengths to get what it wanted in the world.

Among Groves's allies in the spring of 1946 was Senator Arthur Vandenberg of Michigan, who attached an amendment to the McMahon bill establishing a Military Liaison Committee to the commission. It would be the perfect perch for Groves. Even if Groves was an unwise choice for a commissioner, he could nevertheless exert his influence on the commission through the liaison committee.

Yet Groves's opposition to the McMahon bill was also the beginning of his downfall. He had emerged from the war a national hero. On September 21, he and his wife were cheered by a crowd of 5,000 people at New York City Hall, and the mayor presented him with a scroll. He received honorary degrees from colleges and universities and was profiled in national magazines. But Groves had made many enemies while running the Manhattan Project, and after the war they were even more jealous of his success. He clashed repeatedly with those inside and outside of government who wanted access to Manhattan Project files. The atomic scientists caricatured him as a warmongering martinet. His brash personality, so effective in building the bomb, was a liability in the more politically complex postwar world. His final efficiency report said that Groves was "an intelligent, aggressive, positive type of man with a fine, analytical mind and great executive ability. His effectiveness is unfortunately lessened somewhat by the fact that he often irritates his associates."

Groves and his political allies succeeded in weakening provisions of the McMahon bill that would have reduced the military's influence over nuclear matters. But the effort seemed to exhaust him. In 1948, he retired from the army, moved to Darien, Connecticut, and went to work for the Remington Rand Corporation. He gave lots of speeches, played lots of golf, and finally went on vacations with his long-suffering wife. He also got Remington Rand interested in a new technology that had hastened several key aspects of the Manhattan Project: electronic computing.

ON AUGUST 1, 1946, President Truman signed the McMahon bill, officially known as the Atomic Energy Act. It established a five-member civilian commission that would take control of what had been the Manhattan Project and oversee a massive expansion of America's nuclear capabilities. Additions to the bill imposed by conservative congressmen guaranteed that the military would continue to have a powerful influence on nuclear affairs. For example, amendments required that the president could direct the commission to deliver fissionable material and authorize the military to produce nuclear weapons. But the atomic scientists succeeded in wresting at least some of the control of nuclear energy away from the military.

Now they focused their full attention on a much more ambitious goal: achieving some sort of international control of nuclear weapons. In a world where national self-interest came first—even when mutual interests served self-interest—the prospects for international control were never great. Yet the appeals of the atomic scientists had effects. A 1946 poll found that 54 percent of Americans favored transforming the United Nations into a "world government with power to control the armed forces of all nations, including the United States." But when the questions were more detailed, such as asking about specific aspects of national sovereignty that would have to be forfeited, public support dwindled.

Within the US Congress and military, support for international control, much less world government, was virtually nonexistent. Within two weeks of Japan's surrender, the air forces sent Groves a list of 15 cities, including Moscow and Leningrad, that could be targets for atomic bombs in the next world war. In a confidential memo written a few months after the war, Groves argued that "if there are to be atomic weapons in the world, we must have the best, the biggest and the most."

In 1946, the US government proposed a plan for the international control of nuclear weapons that was very unlikely to be acceptable to the Soviet Union. Calling for total disarmament, it would have prohibited the Soviet Union or any other country from developing nuclear weapons. Only after the Soviets and other countries agreed never to build atomic bombs and to open their countries for inspections would the United States dismantle its own weapons. As expected, the Soviets rejected the U.S. plan and made a counterproposal—the United States should eliminate its weapons and then the Soviet Union would consider a system of international control and inspection. The competition between the two superpowers was sharpening. In March 1946, Churchill said that an Iron Curtain was dropping across Europe. In July of that year, just as the negotiations on international control were entering a critical phase, the United States tested two Nagasaki-type bombs near Bikini Atoll in the Pacific Ocean, which the Soviets interpreted as intimidation.

A spy scandal a few months later made it clear that the Soviets had been stealing secrets from the US nuclear program all along.

Meanwhile, the campaign of fear waged by the atomic scientists had backfired. Rather than encouraging Americans to accept international control of nuclear weapons, it made them fear and distrust the Soviet Union and want to protect themselves through any means possible. At the time, the United States was the only country in the world with nuclear weapons. If it could maintain its advantage over other countries, it would not need to rely on international control.

In this climate of heightening international tension, President Truman announced his appointments to the Atomic Energy Commission in October 1946. The head of the commission was David Lilienthal, former head of the Tennessee Valley Authority, which had brought electricity, modern farming practices, and educational programs to the Tennessee Valley in the 1930s. A lawyer, friend of Robert Oppenheimer, and advisor to the federal government on atomic energy, Lilienthal was enthusiastic about using nuclear power for electricity generation and other peaceful purposes. But in the grim international climate of 1947, he knew that the commission would need to focus on military needs first.

Early that year, the commissioners toured the Manhattan Project sites that they now controlled. They were shocked at what they found. The United States had just a single operable atomic bomb. The parts for a few more Nagasaki-type bombs were available, but they would have to be assembled. If the United States wanted to maintain a credible nuclear capacity, work on the bombs had to expand rapidly. And the greatest single need identified by the commissioners was for more plutonium to fuel those bombs.

After almost two years of uncertainty, Hanford was back in business. The Atomic Energy Commission desperately needed what only Hanford could make.

Chapter 20

BUILDING THE NUCLEAR ARSENAL

ON SEPTEMBER 1, 1949, A B-29 OUTFITTED WITH SPECIAL AIR FILTERS took off from an Air Force base in northern Japan and flew to a base near Fairbanks. When it landed, technicians removed the filters and tested them for radioactivity. The tests were positive. More flights crisscrossed the northern Pacific to gather air samples. Within a few days, the evidence was unassailable. The Soviet Union had exploded its first atomic bomb.

At first, Truman refused to believe it. US intelligence agencies had agreed with Groves that the Soviets could not possibly build a bomb so quickly. The statement issued by Truman on September 23 tried to downplay the significance of the event, saying that such a development was inevitable. But for the American public, the shock was enormous.

The bomb exploded by the Soviets on August 29, 1949, in what is today Kazakhstan, was essentially a replica of the plutonium-based bomb dropped on Nagasaki. When Stalin learned about the bombing of Hiroshima, he was furious that Soviet scientists had made so little progress in their wartime atomic bomb program. A few days later, he told his officials in charge of munitions, "Provide us with atomic weapons in the shortest possible time. You know that Hiroshima has shaken the whole world. The equilibrium has been destroyed. Provide the bomb—it will remove a great danger from us."

Soviet scientists and engineers did not need to start from scratch. The Soviet Union had infiltrated the Manhattan Project with spies who kept them thoroughly up to date on the building of the bombs. In the Los Alamos laboratory, the German émigré Klaus Fuchs and at least two other spies provided documents to Soviet agents during and after the war. Several Met Lab employees were suspected of passing secrets to the Soviets and were fired from their jobs. According to documents made public after the collapse of the Soviet Union, the Soviets had a spy codenamed MAR at Hanford—though the details of this espionage are still not known. With the information from spies and from public reports published after the war, Soviet scientists were able to move quickly.

The postwar Soviet Union did not have the industrial might to separate uranium-235 from uranium ore. It therefore pursued only one route to a bomb—the route pioneered by Hanford. Using gulag labor and nearby lakes and rivers for both cooling water and disposing of waste, the Soviets built graphite reactors and reprocessing facilities at the Maiak plutonium plant a thousand miles east of Moscow. The first reactor the Soviets built, to test the purity of graphite and uranium, was a close copy of a reactor built at Hanford for the same purpose. The first Soviet weapon, which was based on the implosion design developed at Los Alamos, even exploded with about the same force as the Nagasaki bomb, lighting up the steppes of central Asia just as the Trinity bomb and Fat Man had lit up the New Mexican desert and the Urakami Valley.

The explosion of Joe-1, as it was code-named in the United States, marked a turning point in the Cold War. A few months later, Truman announced that the United States would embark on the development of hydrogen bombs much more powerful than the atomic bombs dropped on Hiroshima and Nagasaki. Less than two years later, the United States successfully exploded the hydrogen bomb Ivy Mike, which had a force of more than 10 million tons of TNT—five hundred times that of the Nagasaki explosion.

The development of hydrogen bombs had major consequences for

Hanford. Hydrogen bombs work by raising isotopes of hydrogen to such high temperatures that they begin to join together to form helium, which releases energy through the same nuclear reactions that power the sun. The only way to achieve such temperatures in a bomb is to set off a conventional fission weapon. At the center of every hydrogen bomb in the world's nuclear arsenals today is a pit of plutonium, sometimes mixed with uranium, surrounded by high explosives. When the explosives go off, they compress the pit to start a chain reaction, which ignites the light elements adjoining the plutonium bomb. Making hydrogen bombs therefore requires making plutonium pits, and at the beginning of the 1950s all the plutonium for US bombs had to come from Hanford.

The 1950s brought other ominous developments. By the end of 1949, Communist forces had driven the nationalists out of mainland China, and Mao had established the People's Republic of China. On June 25, 1950, North Korea, confident of backing from the Soviet Union and China, invaded South Korea. Also that year, Fuchs, relocated after the war to Britain, publicly confessed that he had passed secret information on the US nuclear program to the Soviet Union, which triggered a panic in the West that other spies had infiltrated the government. During a speech in West Virginia, Wisconsin Senator Joseph McCarthy took a piece of paper from his pocket that he said contained a list of Communists working in the State Department.

The mood in America darkened. Two-thirds of Americans said in a poll that the United States should drop bombs on Russia first in any future war with the Soviets. More than half favored dropping atomic bombs on "military targets" in Korea to bring the war to an end. Some advocated a genocidal preemptive strike that would kill many millions of Soviet citizens and destroy their cities before the Soviet Union could fully arm itself with atomic bombs. Talk of international control had faded away and was replaced only by a desperate need to stay ahead. "In the brutal and strident climate of the early Cold War, hope shriveled," wrote historian Paul Boyer. "What remained was fear—muted, throbbing, only half acknowledged—

and a dull sense of grim inevitability as humankind stumbled toward the nothingness that almost surely lay somewhere down the road—no one knew how far."

DEL BALLARD MOVED to Richland in 1951 to go to work as a Hanford engineer. He had grown up on a dryland farm northwest of Billings, Montana, and a childhood spent building things led naturally to a civil engineering degree at Montana State College in Bozeman. Ballard had two older brothers who had been in World War II, one serving in the Philippines when the atomic bombs were dropped on Japan. When Ballard's brother returned from the Pacific, he told his family he thought he would have been in an invasion force if not for the bombs. When Ballard was in college during the Korean War, he tried to enlist in the air force but was rejected for bad eyesight. Rather than entering the military through the draft, he took a job at Hanford to contribute to the war.

He arrived in Richland on July 3, 1951, moving into ramshackle barracks north of town. It was 105 degrees the next day. "I thought, my lord, what have I gotten myself into?" he recalled years later. "But I was young and willing to accept anything that came my way."

His first job was to inspect the shielding for a new reactor, the C Reactor, that was being built right next to the B Reactor. By the time Ballard arrived in 1951, Hanford was a much different place than it had been just a few years earlier. When the Atomic Energy Commission decided that it needed to produce more plutonium, it quickly realized that the three existing reactors were not going to be enough. It therefore began to build new reactors. The H Reactor, which became operational in October 1949, sat on the shores of the Columbia midway between the D and F Reactors. The next reactor—the fifth after the B, D, F, and H Reactors—broke with DuPont's spacing scheme. The D Reactor, built during the war, had developed problems. Subjected to a fierce bombardment of neutrons, the graphite in the core was swelling so badly that the process tubes were bending, causing the fuel slugs to get stuck inside the reactor. The Atomic Energy Commission decided to build the DR Reactor—for D replacement—next to the D Reactor, thinking that the new reactor could take over when the original one failed. Instead, Hanford's scientists and engineers figured out a way to solve the problem. Counterintuitively, if the reactors were run at higher temperatures and then allowed to cool, the swelling of the graphite blocks subsided. Once the solution was implemented, both the D and DR Reactors churned out plutonium for the next two decades.

DuPont was gone by then. From the beginning, the company's leaders had insisted that they did not want to operate the plant after the war, and the successful conclusion of the war had not changed their minds. DuPont had originally contracted to build and operate the plant for one dollar so that the company would not be seen as a war profiteer, but after the war a government accountant noted that construction had taken only two years rather than the anticipated three. The company therefore received a check for just 68 cents. A few months later, 32 members of the Pasco Kiwanis Club donated a penny each and sent the proceeds to DuPont's president to make up the balance.

In the fall of 1946, the General Electric Company took over Hanford. As with DuPont, General Electric had to be persuaded to take on the job, but Groves wore the company down. Charged by the Atomic Energy Commission with building three new reactors to produce plutonium, General Electric basically copied the designs developed by DuPont and the Met Lab. As with the B, D, and F Reactors, the new H, DR, and C Reactors pumped water out of the Columbia River, through aluminum tubes to cool the chain-reacting slugs, and then back into the river. But they operated at higher power levels than had the three wartime reactors, which meant that more water had to move faster through the process tubes to keep the reactor cool. As a result, more radioactivity entered the river, and the temperature of the river slowly rose as more of it passed through the reactors' cores.

With more reactors operating, General Electric also needed to increase Hanford's chemical processing capacity. At the Met Lab during the war, Seaborg and the other chemists had started thinking about a different way of separating plutonium from the irradiated fuel slugs. Known as redox—a contraction of the term *reduction-oxidation*—the process used a highly flammable solvent to draw plutonium and uranium away from the dissolved slugs. Perfected after the war, solvent extraction was at the heart of a fourth gigantic canyon building erected by General Electric on the central plateau. The solvent extraction done at the REDOX Plant was much more efficient than the batch processing that had occurred in the earlier canyon buildings. But the wastes it generated were even more toxic and radioactive than the earlier wastes. General Electric dumped the less radioactive wastes into the soil near the canyon buildings, as DuPont had done during the war. For the most toxic and radioactive chemical wastes, it built 30 new single-shell tanks and began to fill them up.

By the 1950s, even the REDOX Plant was not enough to meet the plutonium production demands, and a fifth canyon building rose on the central plateau. Known as the PUREX Plant—for plutonium-uranium extraction—it went online in January 1956. Dutifully, General Electric built more single-shelled tank farms to hold the wastes until someone could figure out what to do with them.

Even the six reactors operating at Hanford by the early 1950s were not enough to produce the plutonium the government demanded. In 1952, the Atomic Energy Commission announced that General Electric would build two new reactors at Hanford. Known as K-West and K-East, these plants were much larger than the first six reactors, operating at energy levels almost eight times the design level of the B Reactor. They also used the heat from the cooling water, before it was dumped back into the river, to heat the reactor buildings. It was the first, albeit modest, use of nuclear energy in the United States for purposes other than warfare.

Even as Hanford was growing, the Atomic Energy Commission

was building new facilities elsewhere to meet the needs of the Cold War. In Ohio, 20 miles northwest of Cincinnati, the Fernald Feed Materials Production Center had been producing uranium fuel elements for Hanford's reactors since 1948. In 1952, a new plant to separate uranium isotopes came online near Paducah, Kentucky, about 40 miles up the Ohio River from Cairo. In the high desert between Idaho Falls and the small town of Arco, scientists, engineers, and technicians at the National Reactor Testing Station built and tested experimental reactors, including the reactors now used in nuclear-powered submarines. The government began building the Pantex Plant near Amarillo in the panhandle of Texas in 1951 to assemble nuclear weapons. It set aside the Nevada Test Site northwest of Las Vegas to test nuclear weapons, detonating more than a thousand bombs there between 1951 and 1992.

Of greater long-term consequence for Hanford, the Atomic Energy Commission decided that it needed another place to produce plutonium. Hanford was within range of Soviet bombers coming over the poles from Siberia, which meant that it could be wiped out in a surprise attack. After a quick review of potential sites, the Atomic Energy Commission announced that it would build two new reactors on the Savannah River near Aiken, South Carolina, which soon were followed by three more. Constructed and run by DuPont, which had decided to get back into the nuclear business, the reactors were moderated by heavy water, not graphite, but nevertheless were devoted entirely to plutonium production.

Still, Hanford led the charge. Built to beat Germany to the bomb, then used to end the war with Japan, Hanford had become an indispensable engine of the Cold War.

●

AFTER BALLARD FINISHED his job inspecting the shields for the C Reactor, he did field engineering in the 300 area just north of Richland. Then he helped oversee the laying up of graphite in the K-West and K-East Reactors, after which he served as project engineer for a facility to test the lattice spacing in graphite-moderated reactors. Even as different contractors replaced General Electric in the 1960s, 1970s, and 1980s, he stayed on at Hanford, designing new projects and overseeing their construction. The only thing that changed was the color of his paychecks.

In the early years, the only way to get to Seattle, Spokane, or Portland—all at least 150 miles away—was on slow, two-lane roads clogged with farm traffic. The residents of Richland, Kennewick, and Pasco therefore tended to rely on each other and on the vast surrounding landscape for entertainment. They formed bridge clubs, Bible study, gardening groups, outing organizations, gun clubs, and all the men's business groups found in small towns across America: Eagles, Rotary, Elks, Kiwanis, and Jaycees (along with their female auxiliaries). Among the first dramatic groups were the Richland Players, followed by the Richland Light Opera Company, the Community Concert Association, and the Mid-Columbia Symphony Guild. People went to drive-in movies, rodeos, baseball games, and picnics. Wages at Hanford were good; people in the area bought boats to fish on the lakes and rivers, pickups to drive into the country, and comfortable homes on large lots.

Right after the war the population of Richland dropped from more than 15,000 to less than 13,000 as wartime employees drifted away. But by the time Ballard arrived in 1951, Hanford's Cold War expansion had sparked a population boom. Richland had more than 23,000 residents, with another 25,000 construction workers and their families in a temporary trailer camp north of town. Kennewick and Pasco also were growing—up to about 10,000 people each. In 1947, a new newspaper decided to call itself the *Tri-City Herald* and began promoting the label. Today, people from outside the Tri-Cities are more likely to use that term than the names of the individual towns.

But the three towns have always remained proudly and defiantly independent. From the beginning, Richland embraced its atomic heritage. Local businesses christened themselves Atomic Cleaners,

Atomic Bowl, Atomic Ale Brewpub, and Fission Chips restaurant. Most of the people who lived in Richland worked in a Hanford facility or were a family member of a Hanford worker, so everyone had that in common. But the town was filled with people from all over the United States who had come to Hanford looking for good jobs—even today, Richland has a reputation as a welcoming place because of its history of people gathering from across the United States and building a community. It contained virtually no old people, no minorities, no one who was poor, and no one who was rich. According to historian Michelle Gerber, postwar Richland had the highest birthrate in the nation, "and maternal deaths, infant deaths, and deaths from other causes confounded national averages by being so far on the low side." By 1948, more than 2,000 babies had been born at Kadlec Hospital—named for an army engineer who suffered a heart attack at Hanford during the war and became the first person to die in the new hospital. Wags suggested that the high birthrate resulted from a lack of social activities. A more likely explanation is that the town was full of young men and women healthy and optimistic enough to uproot themselves from elsewhere and start a new life in a forbidding and empty landscape.

Kennewick, centered about 10 miles down the Columbia River from Richland, always had a very different feel. Originally a service town for the surrounding agricultural community, it swelled with overflow Hanford workers and with people who provided services for Hanford workers—shopkeepers, mechanics, barbers, accountants. Kennewick was always the shopping center for the region—no one was surprised when the first shopping mall in the region went up there in later years. Well into the 1960s, Kennewick was a sundown town, where African Americans had to leave before nightfall to avoid run-ins with the police and residents. Jack Tanner, regional director of the NAACP in the Northwest, once called Kennewick "the Birmingham of Washington." But Richland, ostensibly less prejudiced, was not much different: in 1950 it had seven Black residents.

The African Americans and Hispanics in the area usually lived in Pasco, the old railroad town across the river from Kennewick. Even there, most minorities lived in East Pasco, on the other side of the railroad tracks from the rest of the town, in shacks or trailers that often lacked access to water or sewer lines. Pasco was always less connected to Hanford than Kennewick or Richland. When the Atomic Energy Commission began pressuring General Electric in the early 1950s to hire more minorities at Hanford, less than a dozen African American clerks and custodians worked for the company (though contractors employed about 250 Blacks who were helping to build new facilities, jobs that would last a few years).

In 1952, Ballard met a woman named Virginia Kelly at a social event organized by the Young Women's Christian Association, and they were married a year later. A few years after that, the federal government and General Electric began privatizing Richland. In the midst of a competition to demonstrate the superiority of capitalism over communism, the presence of a government-owned and -operated town, even if it was occupied with manufacturing weapons of war, was untenable. Ballard and his wife bought one of the first private lots offered for sale in Richland in 1957. The next year they constructed the third privately built house in Richland, not including the few prewar houses not demolished during the Manhattan Project.

They settled down and began raising a family during one of the most contradictory decades in American history. The parents of the baby boom generation had children while many of them were building, or at least were considering building, fallout shelters in which to survive the end of the world—the Ballards' next-door neighbor had one in his front yard. On their black-and-white televisions, families watched images of mushroom clouds rising from the Nevada Test Site followed by the reassuring fare of *I Love Lucy* and *Leave It to Beaver*. The builder of the atomic bomb, Robert Oppenheimer, lost his security clearance amidst accusations of disloyalty to the United States after he objected, on moral grounds, to the development of the hydrogen bomb. Scientists at the University of California, Berkeley, and other state institutions had to sign loyalty oaths to the United

States or be fired from their jobs. It was a time that combined inno-cence, insecurity, and injustice in equal measure.

Ballard and the other workers at Hanford lived at the heart of this contradiction. President Eisenhower and Soviet leader Nikita Khrushchev, after Stalin's death in 1953, were overseeing enormous expansions of their nuclear arsenals. Both Hanford and the Maiak plant in the Soviet Union maintained a frenzied schedule of construc-tion and plutonium production. The Tri-Cities were ringed by Nike missile sites to shoot down Soviet bombers. Security was incredi-bly tight. Hanford workers were prohibited from talking about their jobs even with many of their coworkers, not to mention their fam-ilies and friends. Levels of plutonium production were top secret, which meant that the amount of radiation being released into the air and water also had to remain secret. And the need to protect workers and the public often conflicted with the need for production.

Yet the residents of Richland considered it "a lovely place to live," as Ballard recalled many years later. "Good schools, peaceful, no crime, good city systems, everything very neat and clean." The pos-sibility of nuclear war "was common in your thinking," he added, but "it was something you didn't really dwell on. . . . We were in a competition to keep our deterrent in place, and that required addi-tional plutonium."

Chapter 21

PEAK PRODUCTION

ON SEPTEMBER 23, 1963, A MARINE HELICOPTER LANDED IN A CLEARED field on the bank of the Columbia River north of Richland. After the dust settled, President John Kennedy emerged from the helicopter surrounded by a gaggle of dignitaries, aides, and Secret Service agents. It was a fiercely hot day—over 90 degrees under a bright and cloudless sky. The men immediately began to sweat in their white shirts, thin dark ties, and wool suits.

Kennedy crossed the field and climbed a dozen stairs to a hastily constructed stage. He looked out at a crowd of more than 30,000 people, the men mostly in short-sleeved white shirts, the women in sundresses and hats. The Hanford nuclear site, normally inaccessible behind high fences, had opened to the public that day so people could hear the president. Schools in Richland were out for Kennedy's visit; high school bands had kept the crowd entertained while people waited for the president to arrive.

Kennedy had come to participate in the dedication ceremonies for the ninth and final plutonium production reactor to be built at Hanford. Known as the New Production Reactor or N Reactor, it had the same basic design as the previous eight reactors—a massive cube of graphite pierced by aluminum process tubes. But the N Reactor had one major innovation. The water flowing through the reactor did not stream back into the Columbia River. Instead, it passed through a heat exchanger, where it generated steam. This steam coursed in pipes over the barbed-wire fences surrounding the

In September 1963, President John Kennedy spoke at the dedication of the final plutonium production reactor built at Hanford. *Courtesy of the Tri-City Herald.*

reactor to a power-generating plant, where it turned turbines and produced electricity. Meanwhile, the cooled water looped back into the reactor to be reheated. When it began generating electricity in 1966, the N Reactor was the largest power reactor in the world and doubled America's nuclear power capacity.

"The atomic age is a dreadful age," Kennedy began, after recognizing the assembled Washington State dignitaries. "No one can speak with certainty about whether we shall be able to control this deadly weapon." But Kennedy did not dwell on the traumas of the previous year, when the United States and Soviet Union almost engaged in full-scale nuclear war over the Soviets' installation of nuclear weapons in Cuba. He was there to express his gratitude to the people of Hanford for their hard work and to tout the peaceful applications of nuclear power. Today, said Kennedy, "we begin work on the largest nuclear power reactor for peaceful purposes in the world, and I take the greatest satisfaction for the United States being second to none." He praised the role played by Washington's powerful congressional delegation in getting the N Reactor approved. He spoke about the need to conserve America's natural resources, to "set aside land and water, recreation, wilderness, and all the rest now, so it will be available to those who come in the future." He urged the state of Washington to use every drop of water in the Columbia River for human benefit, "to make sure that nothing runs to the ocean unused and wasted." Then Kennedy, who would be assassinated just two months later in Dallas, took a pointer tipped with a piece of uranium from the B Reactor, lowered it toward a wildly clicking Geiger counter, and activated a nearby clamshell loader, which dropped a load of dirt into an empty field. "I assume this is wholly on the level and there is no one over there working it," he joked to a nearby dignitary.

In the crowd, three-year-old Kathleen Dillon, who would eventually become a Washington State poet laureate, was sitting on her father's shoulders to watch the president. Many years later she wrote about the experience in a poem entitled "My Earliest Memory Preserved on Film":

> . . . Today the wind is at your back, like a blessing.
> Our long-dead senators applaud
> As you touch a uranium-tipped baton to a circuit
> and activate a shovel atomically.
> This is the future.
> Dad holds me up to see it coming.

◉

BY THE MID-1960S, the postwar fears of the atomic scientists had been realized. The United States had more than 30,000 nuclear warheads that it could use against the Soviet Union and its allies. The Soviet Union had far fewer nuclear bombs—about 6,000—but it was increasing the number rapidly after what it saw as the humiliation of the Cuban missile crisis. Within a few years, the total number of warheads in the world would exceed 65,000, representing almost a million times the destructive power of the bomb dropped on Naga-saki. As is the case today, the warheads were stacked in the weapons depots of military bases around the world, ready to be dropped by long-range and tactical bombers. They tipped intercontinental bal-listic missiles in underground silos, waiting to be launched at the first sign of an incoming attack. They lurked in submarines cruising the world's oceans, a final unassailable reserve with which to wreak vengeance after an exchange of land-based missiles.

By the time of the Kennedy administration, it had become clear that a nuclear war could not be won. Any use of nuclear weapons would likely trigger a large nuclear response from the other side. And even after absorbing a large nuclear attack, either the United States or Soviet Union would have enough weapons left to destroy the other. This standoff came to be known as mutually assured destruction, or MAD. Kennedy's defense secretary, Robert McNamara, described the situation in a pivotal 1965 speech. The security of the United States, he said, depends on its ability "to absorb the total weight of nuclear attack on our country—on our retaliatory forces, on our command and control apparatus, on our industrial capacity, on our cities, and on our population—and still be capable of damaging the aggressor to the point that his society would be simply no longer via-ble in twentieth-century terms. That is what deterrence of nuclear aggression means. It means the certainty of suicide to the aggressor, not merely to his military forces but to his society as a whole."

Even as they were building up their arsenals in the 1950s and 1960s, the United States and Soviet Union—along with other coun-tries working to build nuclear weapons—were seeking to develop peaceful uses of nuclear energy, in part to offset the terror induced by their growing nuclear arsenals. Potential uses included research, medicine, and industrial applications—all of which would be real-ized in the years ahead—but the application that got the most attention was electrical generation.

In 1954, Soviet scientists in the town of Obninsk used an exper-imental graphite-moderated reactor to generate electricity and feed a trickle of current into the grid. On December 2, 1957—fifteen years to the day after Fermi's pile at the University of Chicago went critical—operators powered up the United States' first commercial nuclear reactor. Located in the town of Shippingport, Pennsylvania, on the east bank of the Ohio River, the reactor was based partly on designs developed for nuclear ships and submarines. The successful startup was a public relations coup for the Atomic Energy Commission, which was otherwise devoting most of its energy to building nuclear weapons. But even then, some of the drawbacks of nuclear energy were becoming apparent. The reactor had to have multiple containment structures and emergency cooling systems in case anything went wrong. Originally budgeted at $47.7 million, it ended up costing $84 million to build.

The Atomic Energy Commission never succeeded in separating the peaceful uses of atomic energy from military uses. Part of the problem is that it oversaw both endeavors, even though it tried to keep them separate in the public's mind. In the 1950s, the AEC sponsored advertisements, school programs, and public relations campaigns to present the peaceful side of atomic energy. In the 1956 book *Our Friend the Atom* from Walt Disney Productions, a genie vows that the atom is "our friend and servant" and that nuclear power will bring peace to the world—even as America's nuclear buildup was at its peak.

Another part of the problem is that the AEC had the job of both promoting and regulating nuclear power generation, a fundamental conflict of interest that would plague the commission for decades. The same problem was occurring elsewhere. In Britain, France, and China—the first three countries after the United States and Soviet Union to build atomic bombs—commercial applications of nuclear power were inevitably tied up with bomb-making activities. In the Soviet Union, the connection was even more explicit. Much of the nuclear power in the Soviet Union came from a type of reactor designed to produce both electricity for the grid and plutonium for weapons. Moderated by graphite, these reactors were based on the ones used to build the Soviets' initial atomic bombs, which in turn were based on Hanford's reactors. Among these Soviet reactors were four built in northern Ukraine a few miles from the town of Chernobyl.

●

ONE OF THE PEOPLE who encouraged Kennedy to personally bless the N Reactor was the man he had chosen to head the Atomic Energy Commission—Glenn Seaborg.

By 1963, Seaborg had lived a life most scientists can only dream about. Even as he was doing research at the Met Lab during the war, Seaborg was beginning to explore the nuclear chemistry that goes on in reactors. The absorption of an additional neutron by plutonium-239, which creates plutonium-240, had caused the crisis that required the development of implosion. Plutonium-240 decays by emitting an alpha particle rather than by converting a neutron to a proton, so it does not transform into an element with 95 protons. But what if plutonium-240 absorbed another neutron in a reactor before it decayed, yielding plutonium-241? Seaborg had reason to believe that plutonium-241 might convert a neutron to a proton, which would produce a new element never seen before. The challenge for an element hunter would be figuring out how to isolate tiny quantities of such an element from a jumble of other highly radioactive atoms.

In the end, discovering elements heavier than plutonium required nothing less than a reorganization of the periodic table. The table of elements mounted on the walls of science classrooms has a row of elements that don't seem to fit. Labeled the lanthanide series, they go from element 57 (lanthanum, which was named after the Greek word meaning to lie hidden) through element 71 (lutetium, which was named after the Latin word for Paris). Their forlorn position in the periodic table has to do with the way electrons occupy spaces around ever larger nuclei. With the first 56 elements, each time a proton is added to a nucleus, the corresponding electron gets added essentially to the outermost surface of the atom. With lanthanum, the situation changes. In that element, the final electron gets added not to the surface of the atom but to its interior, in the midst of the cloud of electrons surrounding the nucleus. Only with element 72, hafnium—derived from the Latin name for Copenhagen, after Neils Bohr's hometown—do additional electrons once again get added on the outside.

At the Met Lab in Chicago, Seaborg began to think that the same thing was happening with the elements heavier than uranium. They seemed to be forming a new series of elements, like the lanthanides, with electrons being added to the interior rather than the exterior of atoms. But where did the new series start? The chemistry he and his colleagues had developed to isolate plutonium suggested that it might start with element 89, actinium (named after the Greek word for beam or ray because of its radioactivity). But that element already had a well-established place in the family of elements. Suggesting otherwise would require the heretical act of modifying the periodic table.

When Seaborg suggested the idea to his colleagues, they scoffed. Wendell Latimer, who had provided the clue Seaborg needed to isolate plutonium in 1941, told him that his scientific reputation would be ruined if he made such a proposal. "Fortunately," Seaborg later quipped, "that was no deterrent because at the time I had no scientific reputation to lose." His proposal that actinium begins a new series like the lanthanides, later known as the actinides, was the breakthrough that would win Seaborg a Nobel Prize in 1951, when he was just 39 years old.

In 1944 he and a handful of colleagues in Chicago began looking for new elements in plutonium that had been bombarded with alpha particles in Berkeley's 60-inch cyclotron. They soon isolated an element with 96 protons, followed shortly thereafter by one with 95. Seaborg was about to announce the discovery of the new elements at a meeting of the American Chemical Society when he was scheduled to appear on a radio program called *The Quiz Kids*. One of the Quiz Kids asked him if any new elements had been discovered. Well, since you asked, Seaborg said, yes, you can tell your teachers that we have discovered two new elements—though he later recalled that the students listening to the show were "not entirely successful in convincing their teachers" that they would have to buy new periodic tables.

As with all new elements, Seaborg and his colleagues now had the right to name their discoveries. Because element 95 was chemically related to element 63, europium, they named it americium, after the continent in which it was discovered. Americium is a dangerously radioactive and toxic element. Americium-241 has a half-life of 433 years, which contributes to the long-lasting hazards of spent nuclear fuel. But americium has also saved countless lives. The most common type of smoke detector contains a very small amount of americium-241. The low-energy alpha particles given off by the element ionize the air inside the detector, and the ionized air maintains an electric current between two electrodes. When smoke enters the detector, it interferes with the current, setting off the alarm.

Element 96 is chemically related to element 64, gadolinium, which was named after a Finnish chemist who had worked on the element. Seaborg and his colleagues decided to name their new element curium, after Pierre and Marie Curie.

After World War II, Seaborg stayed at the Met Lab for a year, working on the chemistry of plutonium, americium, and curium. He then returned to Berkeley as a full professor with as much research funding as he would ever need. Over the next 12 years at Berkeley, he and his colleagues discovered six more elements. Element 97, which they also produced by bombarding heavy elements in Lawrence's cyclotrons, is chemically analogous to terbium—a name loosely derived from a town in Sweden. Seaborg and his associates therefore named their new element berkelium, though when Seaborg called Berkeley's mayor to tell him that his town would be immortalized in the periodic table, the mayor greeted the news "with a complete lack of interest."

The chemical equivalent of element 98 is dysprosium, from a Greek word meaning hard to get. This did not suggest an appropriate name, so Seaborg and his colleagues named element 98 californium. *The New Yorker*'s "Talk of the Town" chided them on April 8, 1950. "While unarguably suited to their place of birth, these names strike us as indicating a surprising lack of public-relations foresight on the part of the university, located, as it is, in a state where publicity has flourished to a degree matched perhaps only by evangelism. California's busy scientists will undoubtedly come up with another atom or two one of these days, and the university might well have anticipated that. Now it has lost forever the chance of immortalizing itself in the atomic table with some such sequence as universitium (97), ofium (98), californium (99), berkelium (100)."

Seaborg and his colleagues wrote back:

"Talk of the Town" has missed the point in their comments on naming of the elements 97 and 98. We may have shown lack of confidence but no lack of foresight in naming the elements "berkelium" and "californium." By using these names first, we have forestalled the appalling possibility that after naming 97 and 98 "universitium" and "ofium," some New Yorker might follow with the discovery of 99 and 100 and apply the names "newium" and "yorkium."

The New Yorker responded: "We are already at work in our office laboratories on 'newium' and 'yorkium.' So far we just have the names."

The discovery of the next two elements came from an unexpected source. When Seaborg and his Berkeley colleagues heard that a new isotope of plutonium had been discovered in the test debris from the United States' first hydrogen bomb, they wondered if the power of the bomb could also produce heavy elements. They acquired from friends working in the weapons program a piece of filter paper from a plane that had collected samples after the blast. From the dust on the paper they identified element 99 and then element 100—discoveries that were made at about the same time by separate groups at laboratories outside Chicago and in Los Alamos. The codiscoverers named element 99 einsteinium and element 100 fermium.

Fermi got the news that his name would be added to the periodic table on his deathbed. After the war he had moved back to the University of Chicago to become a professor, though he continued to advise the government on its nuclear weapons program. In the summer of 1954, Fermi returned to Europe for just the second time since receiving the Nobel Prize in 1938. While giving a series of lectures and hiking through the Alps and Dolomites with friends, he began to have troubles with his digestion. He and Laura ascribed it simply to the stress of recent years, but his friends were troubled by how thin he looked. Back in Chicago, he went to a doctor, who told him that the cause was psychological, which he doubted. By October, doctors began to suspect something more serious. Exploratory surgery revealed metastatic stomach cancer. Nothing could be done.

Laura rented a hospital bed so he could spend his remaining time at home. Enrico told her to rent it only until the end of November, since he wouldn't need it after that. In his last few weeks, Leona Woods visited often. She, too, had moved back to Chicago after the war to continue working with Fermi. In his final days, they talked about writing one more paper together. He joked that his name would have to be followed by a black cross directing the reader to a footnote saying, "Care of St. Peter." After each visit, Woods drove home in tears.

Fermi died on November 29, 1954. Because his illness was so sudden, he never had a chance to write his memoirs or otherwise reflect on the role his discoveries play in the development of atomic bombs. Then again, he never betrayed much emotion about the implications of his work, beyond the excitement he felt in discovering new things. The science came first.

Fermi was just 53 when he died. Did his frequent exposures to radiation have anything to do with his early death? It's possible, but it can't be determined with certainty. None of his colleagues from Italy, with whom he had done his early work on radioactive substances, died from cancer. Almost 5,000 people under the age of 55 are diagnosed with stomach cancer every year in the United States, very few of whom have worked around radioactivity.* Fermi may just have gotten unlucky. But questions about radiation exposures and health would be asked with increasing frequency—and with increasing vehemence—in the years ahead.

○

IN ADDITION TO DISCOVERING new elements, Seaborg became an accomplished administrator in the years after World War II. He served on the General Advisory Committee of the Atomic Energy Commission during the 1940s and 1950s, advising the government on the expansion of the US nuclear weapons program. At the same time, he was rising up the academic ladder at the University of California, Berkeley. First he became faculty athletic representative, a job he enjoyed immensely because of a lifelong interest in sports. Then he became head of the university, which he later described as "the most difficult job I ever had." Given his accomplishments, he must not have been terribly surprised when President-Elect Kennedy called him on January 9, 1961, to ask if he would chair the Atomic Energy Commission. That evening, he asked Helen and their six children whether they wanted to move from California to Washington, DC. They unanimously said no, they all wanted to stay in Berkeley (though Seaborg later wrote that he had "doubts about the validity of fourteen-month-old Dianne's vote"). Still, he overruled them, figuring that a few years in the nation's capital would be a good experience for them all.

Seaborg never gave up on plutonium. For its first two decades on Earth, it had been used almost exclusively to fuel nuclear weapons. But Seaborg always believed that it was destined for great things. By converting uranium-238 into plutonium-239, nuclear reactors can

---

* My grandfather, who spent the latter part of his life as a farmer downwind from Hanford, also died of stomach cancer, but exposure to radioactivity was almost certainly not the cause of his illness.

unlock all the potential nuclear energy in uranium ore, not just the energy available in uranium-235. Once plutonium was generated in a reactor, reprocessing plants like the ones at Hanford could extract it for use in other reactors, providing humanity with essentially unlimited amounts of energy. But this bounty would come at a large cost. Producing plutonium for nuclear energy would require a huge new industry, which some have called the "plutonium economy." This industry would inevitably create large stores of plutonium, which could be diverted to nuclear weapons programs. And reprocessing would produce vast quantities of toxic and radioactive chemicals, which would need to be isolated from the environment for many thousands of years. Still, Hanford had demonstrated that at least the first part of a plutonium economy would work—reactors could convert uranium to plutonium, and this plutonium could be extracted from spent fuel. And in the first half of the 1960s, with science and technology ascendant, with the United States working to send men to the moon, anything seemed possible.

On June 7, 1968, Seaborg returned to Richland to celebrate the 25th anniversary of Hanford. With Groves and Matthias both in attendance at his speech, Seaborg termed plutonium "the fuel of the future." He predicted that it would power spacecraft, which it subsequently did, and also artificial hearts, which it did not. He joked that plutonium's value "may someday make it a logical contender to replace gold as the standard of our monetary system." He observed that nuclear energy had transformed the area around Hanford. "The atom has been responsible for some healthy and happy communities, and Richland is certainly among the foremost of them."

In his speech in Richland, Seaborg said that plutonium was going to create energy "on such a grand scale and so cheaply" that it "will radically change our relationship to almost all other materials," including food, water, air, and minerals. Nuclear power "is being accepted for economic reasons," he said. "It is being accepted for environmental reasons. And it is being accepted for aesthetic as well as for practical reasons. The time is not far off when clean, compact, competitive nuclear power plants will be the 'conventional' power plants of the day." Nuclear plants would desalinate seawater, produce gasoline from coal, and generate steam heat for factories. Heavy industry could be located around large nuclear power plants, creating what Seaborg called a Nuplex. Such facilities could pump water from deep underground, create new and exotic materials, and even economically recycle wastes, which is "increasingly important when the disposal of our output of junk and garbage is becoming a problem of major proportions." The Nuplex would physically separate heavy industry from cities "so that we could plan and build cities designed for the ultimate in healthy living. Powered by the Nuplex, cities like Richland could become urban utopias.

In 1968, Seaborg was speaking during the peak period of enthusiasm for commercial nuclear power. Over the previous two years, US utilities had ordered 67 nuclear reactors. By the time Seaborg left the Atomic Energy Commission in 1971, 25 reactors were operating, 52 were under construction or being reviewed for operating licenses, and 39 were being reviewed for construction permits. At that point, nuclear energy was generating about a fifth of the United States' electricity, as it does today.

"Perhaps 25 years from now we will be able to gather here to look back over half a century of progress of the Nuclear Age," Seaborg predicted. "By then Richland, together with the Tri-Cities Area, will probably be a large metropolis thriving on its growing science-based industries. Perhaps Hanford will be its Nuplex, able to preserve the surrounding vast and majestic area close to the way nature created it. And we will be able to reminisce about the beginning of the Nuclear Age while we see all about us many of the wonders that it has brought and continues to unfold."

○

MANY OF THE PEOPLE listening to Seaborg's speech that day in Richland were desperate for such a vision. Since Kennedy's visit five years before, Hanford's prospects had nose-dived. The reason was simple: after two decades of plutonium production at Hanford and South Carolina, the United States had all the plutonium for weapons it would ever need. By 1968, the reactors on the Columbia and Savannah rivers had produced 90 metric tons of weapons-grade plutonium—enough for more than 14,500 Nagasaki-type bombs. Newer bombs used even smaller plutonium pits, which meant that more could be built.

Political considerations also reduced the need for plutonium. In the early 1960s, Seaborg had helped lead the negotiations on the Comprehensive Test Ban Treaty, which banned all testing of nuclear weapons in the atmosphere and outer space. The ban was not really comprehensive, since testing could still occur underground, which led to continued rapid development of nuclear weapons. But with live images of mushroom clouds removed from the evening news, people were less fearful of nuclear war, and the need to intimidate the Soviets seemed less pressing. As a result, the US military could shift funding from Cold War armaments to the very hot and escalating war in Vietnam.

Ironically, the first reactor to close was the replacement for the D Reactor, in 1964. The H and F Reactors closed the next year, and the D, B, and C Reactors closed in 1967, 1968, and 1969. By 1971, after the shutdown of K-West and K-East, the only reactor still running was the N Reactor, and even it was no longer producing weapons-grade plutonium. The reactors that the United States had built to end World War II and help win the Cold War sat unused and abandoned on the desert floor.

○

KATHLEEN DILLON, who later wrote the poem about seeing John Kennedy in 1963, never knew a time when she was not best friends with Carolyn Deen. Born a few months apart in Richland's Kadlec Hospital in 1960, the girls lived in identical Y houses three doors away from each other near the intersection of Cedar Avenue and Cottonwood Drive. Every day they walked to and from Marcus Whitman

Elementary School, and then Carmichael Junior High School and Richland High School, sharing both the trivia and the turning points of their lives. "I learned how to be a friend from Carolyn," Kathleen later said. This was a time in small town America when children did not know that the doors of their homes could lock, when car keys were left in ignitions so they wouldn't be misplaced, when children of any age could go wherever they wanted so long as they were home for dinner. Each of the mothers had her own signal—a whistle, a cowbell, a ranch-style triangle—to indicate that it was time to eat. If a couple of extra kids showed up for a meal, there always seemed to be enough food to go around.

Carolyn's father, Tom, came to Hanford in 1951 when he was 22 years old. He had just finished serving in the Korean War, had a new wife, and had heard that men with a high school diploma could find good jobs at Hanford. For the next 27 years, he did the kinds of things that thousands of other Hanford workers did. He canned fuel elements on the production line, worked a metal lathe, drove a high lift to load and unload uranium billets. He rarely if ever got sick and often volunteered for special jobs to make extra money. He chopped up railroad boxcars that had been contaminated and bundled up the radioactive planks for disposal, often with little or no protective clothing. Sometimes he served as a jumper—a worker who would enter a radioactive area, make a repair, and then quickly leave, having received a month's permissible radiation dose in just a few minutes. He was a union member and served as union president of the Hanford Atomic Metal Trades Council. Toward the end of his career, he became manager of labor relations for UNC Nuclear Industries, which operated the reactors at Hanford.

Kathleen's father, Robert, was a chemist from Portland with a PhD from Northwestern University. When he got his doctorate, he received two job offers—one from Texas, one from Hanford. He chose the offer that was closer to his childhood home. He spent his career at Hanford studying aluminum corrosion to better understand what happens inside a reactor's process tubes, working for much of that time in an office reportedly used by Fermi. Research could be done at Hanford that was impossible to do elsewhere. Scientists working at laboratories associated with Hanford developed new methods of monitoring radiation exposures. They studied the effects of radiation on fish, plants, livestock, and humans. They investigated how radiation affects materials. By the 1960s, thousands of people were working at the Pacific Northwest Laboratory at Hanford, a research enterprise run by the Ohio-based Battelle Memorial Institute.

In the 1960s the Tri-Cities and the towns surrounding it were isolated, proud, and self-reliant. Directly or indirectly, most people relied on the federal government for their jobs, and people supported the government and its leaders. As Carolyn recalled much later, "If you were going to work here, you'd better believe in what the government was doing." Teenagers enlisted or were drafted into the military and went to Vietnam, but the war was far away, and the protests that erupted in other parts of the United States were largely absent in the Tri-Cities. For the most part, it was an innocent and largely carefree place and time to grow up in America.

But growing up in an atomic city had its peculiarities. When Carolyn and Kathleen were in elementary school, a semi-truck once pulled up outside the building with the words "Whole Body Counter—Mobile Laboratory" printed on the side. One by one, the students filed into the truck and lay down on a flat rubber platform. A motor whirred, and the platform moved the recumbent child through a large cylindrical radiation detector. As an adult, Kathleen wrote a poem about the experience:

> We were told to close our eyes.
> Everyone was school age now, our
> kindergarten teacher reminded us,
> old enough to follow directions
> and do a little for our country.
> My turn came and the scientists
> strapped me in and a steady voice

prompted, The counter won't hurt,
lie perfectly still, and mostly I did. . . .
Just once I peeked
and the machine had taken me in
like a spaceship and I moved
slow as the sun through the chamber's
smooth steel sky.

Kathleen and Carolyn were children in Richland during the years of peak plutonium production—their monitoring in the whole-body counter was part of an experiment to see if emissions from the plant were getting into the food and water of people in the area. But they were also children when production was ramping down. In 1968, when they were in the third grade, the principal of Marcus Whitman Elementary School came into their classroom and told them that they were going to write letters to President Johnson pleading for the K Reactors to stay open. Later, Kathleen was one of the students chosen to accompany the bags of mail to the Pasco airport for delivery to Washington, DC.

The letter-writing campaign failed—K-West and K-East both closed a few years later. Yet employment at Hanford did not go down much, even after eight of its nine reactors were closed. Washington State's congressional delegation was so influential that it was able to keep funds flowing to the site—plant managers and union leaders found things for people to do.* But the government largess would not last forever. Hanford would have to change to remain a viable operation. What would it become?

---

* My other grandfather was a steamfitter who lived in a trailer with my grandmother and travelled throughout the western United States for much of the second half of his life. He occasionally worked for a few months at Hanford during the period when union pipefitters there operated by the motto *"nobody can lay pipe too slowly."* He reported that he played a lot of gin rummy.

Chapter 22

THE RECKONING

BY THE LATE 1970S, SEABORG'S VISION OF A NUPLEX RISING FROM the desert outside Richland seemed to be coming true. Earlier that decade, a consortium of publicly owned utilities called the Washington Public Power Supply System announced plans to build three nuclear power reactors on the Hanford reservation, and by the early 1980s swarms of construction workers were pouring foundations and laying pipe. Boosters in the Tri-Cities talked of 15 to 20 nuclear plants being built just upriver, more than replacing the shutdown production reactors.

Even plutonium production was about to return, though in a limited way. In the 1970s, a bipartisan coalition of US foreign policy hawks, loosely affiliated under an organization called the Committee on the Present Danger, began to issue warnings of a strategic imbalance between the United States and the Soviet Union. Their arguments never made much sense. By that time, each nation had approximately 25,000 nuclear warheads. Given estimates made about the same time that, at most, 300 atomic bombs dropped on either the United States or Soviet Union would end each nation's existence, the antagonists had almost a hundred times more warheads than they needed.

Advocates' claims of a missile gap grew from political expediency, naked self-interest, and, to some extent, legitimate concern about the actions and intentions of the Soviet Union. Hard-line politicians like Scoop Jackson from Washington State and his assistant

The N Reactor, in the foreground, produced plutonium for nuclear weapons and steam for electricity from 1966 through 1987. The K-West and K-East Reactors (emitting steam) and the B Reactor (emitting a plume of dark smoke) are farther up the Columbia River. *Courtesy of the US Department of Energy.*

Richard Perle, who later became a prominent foreign policy hard-liner, played the same card Groves had played during the Manhattan Project: in the face of uncertainty, assume the worst of an enemy and act accordingly. In the late 1970s, under the leadership of an aging Leonid Brezhnev, the Soviet Union was again behaving in threatening ways. Earlier in the decade, Cuba had become involved in a civil war in Angola with Soviet backing. Soviet support for governments in the Middle East was heightening tensions in that key geopolitical region. Then, in December 1979, the Soviet Union invaded Afghanistan. In the following US election year, which pitted the conservative former governor of California, Ronald Reagan, against the incumbent, Jimmy Carter, stoking fears of Soviet nuclear supremacy was an obvious way to get votes.

The N Reactor had been chugging along since 1966, mostly to

produce steam for the electricity-generating plant next to it. But it had not been producing weapons-grade plutonium because no one foresaw a need for more after the production frenzy of the 1950s and 1960s, and the South Carolina reactors were newer and better suited to make plutonium. (The South Carolina reactors also made tritium for the United States' hydrogen bombs.) Nevertheless, bowing to political pressures, both the Carter and then the Reagan administrations called for the N Reactor to begin producing more plutonium for bombs. After being upgraded, it began producing weapons-grade plutonium in 1982. The next year, the PUREX Plant, after an 11-year shutdown, resumed separating plutonium from irradiated uranium.

After years of declining employment at Hanford, the Tri-Cities seemed on the verge of a turnaround. New reactors were going up north of Richland. Hanford was producing more plutonium. During the 1970s, the federal government had built an experimental reactor at Hanford, the Fast Flux Test Facility, to evaluate new commercial reactor designs, and the reactor began operating in 1980. Despite a severe recession in the United States at the beginning of the 1980s, Hanford's future seemed bright.

Then it all came apart.

●

ON JUNE 21, 1982, Jim Stoffels, a physicist at the Pacific Northwest Laboratory, saw a notice on page five of the *Tri-City Herald*. "The Tri-Cities Nuclear Weapons Freeze Campaign is to meet at 7:30 p.m. Thursday at Sixth and Clark, Pasco," it said. The nuclear freeze movement was a worldwide reaction against the heightened militarism of the late 1970s and early 1980s. It called on the United States and Soviet Union to stop testing, producing, and deploying nuclear weapons, both to reduce international tensions and to halt the arms race. With support from prominent public figures and scientists, the idea caught fire with the American public. A week before Stoffels saw the notice in the *Tri-City Herald*, a million people congregated in New York City to demonstrate in favor of a nuclear freeze—still the largest single demonstration in US history. In the fall elections, more than 60 percent of voters supported referenda calling for a freeze, and the movement helped Democrats pick up 27 seats in Congress.

Stoffels had been working at the Pacific Northwest Laboratory near Richland for nearly two decades by then, after graduating from Marquette University with a master's degree in physics. He was part of a group using a technique called mass spectroscopy to measure levels of uranium, plutonium, and radioiodine in the environment. Tall, thin, with wavy brown hair, Stoffels was a deeply religious man who always sought to apply the tenets of his Catholic faith in his daily life, but he had not previously been politically active on nuclear weapons.

Within a few days of the meeting, Stoffels had become founding vice president of a new organization, World Citizens for Peace–Tri-Cities. A month later, World Citizens for Peace held its first public event—a commemoration of the atomic bombings of Japan that drew about 100 people to John Dam Plaza across the street from Richland's federal building. "No one expected a big crowd for a nuclear freeze gathering in Richland," reported the *Tri-City Herald* the next day. "But some were encouraged that such an event would occur at all."

Later that year, the chair of World Citizens for Peace, Maurice Warner, was asked by his boss at the Pacific Northwest Laboratory to lead an environmental assessment of the proposed MX missile "rail garrison" system, which would have carried armed intercontinental ballistic missiles around the United States on railcars so that they would be harder for the Soviet Union to target. Warner, a Quaker, refused the assignment, resigned from his job, and moved to Seattle to become a career counselor. By January 1983, Stoffels was the new chair of the group.

World Citizens for Peace began holding monthly peace vigils in different locations around the Tri-Cities. A particular focus of their attention was the White Train, which carried nuclear weapons from the Pantex assembly plant in Texas to military bases around the country. One such destination was the Bangor submarine base 20 miles west of Seattle, which is the West Coast base of America's Trident submarines and has one of the largest concentrations of nuclear weapons anywhere in the country. The military painted the railcars white to keep the weapons cool, but that made them an obvious target as the White Train wound its way through the Tri-Cities. On March 21, 1983, World Citizens for Peace held a candlelight vigil at the Pasco depot as a train believed to hold 100 nuclear weapons rolled by. Not long thereafter, the federal government painted the cars different colors to escape detection and subsequently used unmarked tractor-trailers to move warheads rather than railcars.

Two months later, World Citizens for Peace held a demonstration outside Hanford's main gate to protest the restart of the PUREX Plant. "Plutonium Production Supports the Insane Nuclear Arms Race," one sign read. "Stop Contaminating the Columbia River," said another. Some people driving by in their cars flashed a thumbs-up as they passed. Others used a different finger, and one employee used his rear end to demonstrate an opinion. Still, Stoffels was surprised and encouraged that so many people in Richland supported the group's message.

As the 40th anniversary of the atomic bombings approached in 1985, World Citizens for Peace adopted "reconciliation" as the permanent theme of its annual memorial ceremony in Richland. That summer, Stoffels wrote to the mayor of Nagasaki about the group's plans for the ceremony, and the mayor wrote back to say that Nagasaki would like to cooperate in the memorial by presenting Richland with a model of the "Bell of Peace." The original bell was recovered from the ruins of the Urakami Cathedral, the mayor wrote, and was rung every day during the city's recovery to console the survivors of the bombing. The mayor pro tem of Richland, Bob Ellis, accepted the gift of the bell on behalf of the city, saying, "The desire for peace is universal in the hearts of mankind." Every August since then, the ringing of the Bell of Peace has been the highlight of the Atomic Cities Peace Memorial ceremony held by World Citizens for Peace.

Even as the nuclear freeze movement was gathering steam, the dreams of a Nuplex at Hanford were fading. On March 28, 1979, Reactor Number 2 at the Three Mile Island Nuclear Generating Station, 10 miles southeast of Harrisburg, Pennsylvania, experienced the most serious nuclear accident in US history. Most power reactors in the United States do not use graphite to slow down neutrons and split atoms of uranium-235. Instead, they use ordinary water and "enriched" uranium, in which the percentage of uranium-235 is higher than its natural 0.7 percent, to compensate for the neutron-absorbing properties of the cooling water. The water in such a reactor not only moderates the reaction but transfers heat from the core of the reactor to the turbines that generate electricity. The problem is that if water stops flowing to the core, the nuclear fuel will heat to the melting point, even with all the control rods inserted into the reactor to absorb neutrons. Commercial power plants therefore need multiple backup systems to make sure their cores never run dry.

At Three Mile Island, a stuck valve, poor design, and human error led the operators to turn off the emergency cooling system even as water was draining away from the core. As the fuel elements were exposed to the air, they began to melt. Eventually the operators figured out what was going on and refilled the core with water, but not before the interior of the reactor was seriously damaged. The accident released only a small amount of radioactivity into the atmosphere, not enough to cause any health effects that could be tied to the accident. But as with the eruption of Mount St. Helens in Washington State the following spring, television reporters had just begun using trucks outfitted with microwave transmitters, and their on-the-scene reports gave the accident a riveting immediacy. Coincidentally, the movie *The China Syndrome* had premiered two weeks before the partial meltdown, and its fictional account of a TV newswoman and her cameraman, played by Jane Fonda and Michael Douglas, reporting from inside a nuclear power plant during a major accident seemed to provide a script for what had happened in Pennsylvania.

Actually, the nuclear power industry was in trouble well before the body blows of Three Mile Island and *The China Syndrome*. Despite their efficiency, large commercial nuclear power plants were very expensive to build and operate, largely because of the safety systems and regulations required to protect the public in case of an accident. By 1979 the environmental movement had turned resolutely against nuclear power, and antinuclear advocates had increasing success emphasizing the link between nuclear power and nuclear weapons.

The combination of antinuclear sentiment, the recession, escalating costs, construction delays, and poor management was too much for the Washington Public Power Supply System. In 1981, it halted construction on two of the five reactors it was building—one in western Washington and one at Hanford. In 1983, it mothballed two more half-finished reactors. Finally, later that year, it defaulted on the billions of dollars of bonds it had sold against the promise of future electricity—still the largest bond failure in US history. In the end, only one reactor was completed, leading people to pronounce the acronym for the organization—WPPSS—as "whoops." Today, the Columbia Generating Station north of Richland continues to produce about 10 percent of the electricity generated in Washington State.

Worse was yet to come. In 1986, plant operators at reactor number four of the Chernobyl Nuclear Power Plant in northern Ukraine tried to conduct a test that involved turning off the plant's safety systems. During the test, the chain reaction got out of control and an explosion in the core tore open the reactor. Commercial power reactors in the United States are required to have heavy steel and concrete containment structures around the core to keep radioactivity from escaping during accidents. Chernobyl had no such containment vessel, partly because its operators thought an accident could not happen and partly because it was designed to produce both electricity and plutonium for weapons, which made a containment vessel inconvenient. When the graphite in the core caught fire from the heat of the nuclear fuel, it burned like a bonfire. Fission products and transuranic elements rose into the atmosphere and settled across broad swaths of Europe.

In 1986 the United States had just a single operating reactor that used graphite as a moderator and generated both electricity and plutonium: the N Reactor, on the right bank of the Columbia River, 30 miles north of Richland. N Reactor was different in several key ways from the Chernobyl reactors, but it, too, like the other military reactors at Hanford, did not have a containment vessel. Denied access to Chernobyl, reporters flocked to Richland to draw comparisons with what was happening in the Soviet Union. A few months later the N Reactor shut down for safety upgrades. It never operated again.

The Chernobyl accident made a powerful impression on the newly installed Soviet general secretary. If a nuclear accident could cause this much physical damage and health risk, thought Mikhail Gorbachev, what would a nuclear war do? "The accident at Chernobyl showed again what an abyss will open if nuclear war befalls mankind," he said in a speech 18 days after the accident. "The stockpiled nuclear arsenals are fraught with thousands upon thousands of disasters far more horrible than the one at Chernobyl."

Even before the Chernobyl accident, Gorbachev, who was desperately trying to reform the Soviet economy and military, had proposed in a letter to Reagan that their two countries "agree on a stage-by-stage program" so that "by the end of 1999 no more nuclear weapons remain on Earth." Now he repeated his proposal. "Global nuclear war can no longer be the continuation of national politics, as it would bring the end of all life, and therefore of all politics," he told the Politburo a few months after Chernobyl.

Gorbachev's proposals resonated with Reagan, despite his earlier belligerence toward the Soviet Union. Since his days as California governor, Reagan had talked about reducing the threat to humanity posed by nuclear weapons. As he said in his 1984 State of the Union address, "The only value in our two nations possessing nuclear weapons is to make sure they will never be used. But then would it not be better to do away with them entirely?" He eagerly accepted Gorbachev's invitation to meet in Reykjavik, Iceland, in October 1986 to discuss arms reduction.

The summit between Reagan and Gorbachev would make a marvelous movie someday, though whether it would be a comedy or a tragedy is uncertain. They met in Hofdi House, the former residence of the French consulate, which sits on a broad expanse of treeless lawn overlooking Reykjavik Harbor. Sometimes they talked just with their translators and notetakers, sometimes with their top aides. Over the course of two days, the two men were in turn belligerent, hopeful, obtuse, and dismissive. Reagan told the same story so often that Gorbachev would block his ears when he saw it coming. Gorbachev let his anger flash when Reagan refused to acknowledge or concede a point. Nevertheless, over the course of their discussions, the two made progress. They agreed to eliminate intermediate-range nuclear missiles from Europe. Gorbachev said that the Soviet Union would reduce the number of its missiles targeting Asia to one hundred if the United States would deploy no more than that in Alaska. Finally, in their last session together, as they were arguing over which categories of missiles should be slated for reductions, Reagan said, "It would be fine with me if we got rid of them all."

"We can do that," Gorbachev replied. "We can eliminate them all."

It was an electrifying moment in world history—the first time the leaders of the world's two nuclear superpowers, meeting face to face, had contemplated ridding the world of nuclear weapons.

But there was a hitch. Reagan had become convinced that a space-based defensive system could be built that would use satellites and interceptors to shoot down missiles kept hidden by the Soviets or launched by a rogue nation. Gorbachev refused to let the United States extend the arms race into space. He proposed that the United

States be allowed to work on the Strategic Defense Initiative—or Star Wars, as it was called after the 1977 George Lucas movie—only in the laboratory, not in orbit around the Earth. Reagan refused to limit the scope of SDI research. The two men could not agree. The opportunity passed.

The idea that some sort of defensive system could shield the United States from nuclear weapons was based on a fundamental misunderstanding. Stopping a massive nuclear attack was no more possible in 1986 than it was the week after Hiroshima and Nagasaki, and the situation is the same today. Just seven years after Reykjavik, the Clinton administration canceled the Strategic Defense Initiative, though work has continued since then on much more limited ground-based defenses.

An agreement by Reagan and Gorbachev to eliminate nuclear weapons probably would not have held. Too many powerful interests opposed to disarmament were arrayed beneath the two leaders. But what if they had emerged from their meeting room in Hofdi House and had announced to the world that they had vowed to destroy the nuclear weapons they controlled? Would we still live in a world where human civilization could be destroyed in a few hours through mishap, malevolence, or madness?

●

AFTER GRADUATING FROM Richland High School in 1978, Kathleen Dillon and Carolyn Deen went in different directions. Though Carolyn was a good student, her family did not have enough money to send her to an expensive college, and she wanted to stay close enough to her parents and brothers to come home on weekends. She decided to go to Eastern Washington University near Spokane and become an elementary school teacher.

Kathleen had not been all that interested in science in high school, but she was always impressed by the scientists her father brought home for dinner. They seemed urbane, educated, experienced—a window on a place far away from Hanford. She decided to go to

Washington State University, on the border of Washington and Idaho 100 miles east of Richland, and study engineering.

In 1983, with a degree in civil engineering, Kathleen moved back to Richland and went to work at Hanford as an environmental engineer. One of her jobs was to perform what was called a water balance. As Hanford dumped contaminated water into the ground, the extra water created underground mounds in the water table, which drove contaminants faster toward the Columbia River. A water balance compared the amount of water being withdrawn from the river with the amount being officially released from the reactors, separation plants, and other facilities. The two never matched up. Kathleen's job was to figure out where the missing releases were taking place.

The old-timers at Hanford were not eager to tell her. Partly, Kathleen later reflected, they were still part of a "need to know" culture that originated in the Manhattan Project. What business was it of hers where the water went? "My water balance probably sounded to them like busy work—and doomed. Everybody knew that the facilities were held together with baling wire and bubble gum. Even if you could get numbers on the instrumented effluent streams (and the instruments were rarely calibrated), what about all the leaks they knew about, and the ones they didn't." It didn't help that she was a 24-year-old woman while they had worked at Hanford for decades. The old-timers knew the plant better than any environmental engineer or government official could know it. They had built and run Hanford, and it had never blown up or caused any obvious problems. On the contrary, it had provided a material that America desperately needed during the darkest days of World War II and the Cold War—along with good jobs and a sense of purpose. If that required burying and then ignoring radioactive wastes in the desert sands, well, that was the price of security.

But by 1983 the problems posed by Hanford's wastes were becoming impossible to ignore. Since 1943, the federal government had built 177 huge underground tanks near the chemical separa-

Immense steel and concrete tanks, shown here under construction, still hold millions of gallons of highly radioactive chemicals produced by plutonium separation. *Courtesy of the US Department of Energy.*

tion plants. The tanks consisted of circular, concrete shells with one and, later, two layers of interior steel linings. They contained the chemical and radiological effluent of 40 years of plutonium production—a toxic mishmash of chemicals, fission products, and unrecovered uranium and plutonium so radioactive that, if held in a glass at arm's length, the waste would deliver a fatal dose in just a few minutes. Buried beneath 6 to 11 feet of dirt to protect people from their radioactive contents, the tanks were always meant to be an interim solution for the separation plants' liquid wastes, but no one had come up with a permanent solution. Meanwhile, the early single-shelled tanks began to leak not long after they were built; a million or so gallons of waste had already flowed into the rocks and soil below the tanks.

In addition to the high-level waste at Hanford, plant operators had released more than 400 billion gallons of water from Hanford's reactors, canyon buildings, and other facilities directly into the ground—nearly a thousand times the volume of waste in the tanks.

They assumed that the chemicals and radioactive elements in the water would bind to the dry desert soil and remain immobilized. But some of the contaminants, carried deeper by rainwater and continued wastewater flows, moved farther downward. Eventually they hit the water table beneath the dry soil. There they began moving sideways, toward the Columbia River.

By the 1980s, something else about Hanford was becoming impossible to ignore. For years, farming families on the flat expanse east of Hanford had been adding up the number of cancers, miscarriages, and stillbirths that they and the people they knew had experienced. The total seemed suspiciously high. Environmentalists had been saying that the radiation released from nuclear facilities and tests was making people sick. Now the farmers—and increasingly the residents of the Tri-Cities—had to ask themselves a question: Could their illnesses, and those of people they knew and loved, be the result of living near Hanford?

It was a hard question for people in the region to confront. They were proud of their contributions to ending World War II and supplying America's nuclear arsenal. Now they had to ask whether the sacrifices they had made were greater than they had known. As books and magazine articles began to appear in the 1980s about Hanford's assaults on human health and the environment, written mostly by outsiders, residents of the region became increasingly concerned—and also defensive. In 1986, a group called the Hanford Family formed to defend the Tri-Cities and its largest employer from what they called "Hanford bashing." Using grassroots methods common among antinuclear groups, they organized volunteers, distributed flyers and newsletters, and held rallies to attract newcomers and spread their message. Former Washington governor Dixy Lee Ray—a lifelong supporter of nuclear power and the last chair of the Atomic Energy Commission before its responsibilities were divided in 1975 between the Department of Energy and the Nuclear Regulatory Commission—spoke to the group wearing a "Proud of Hanford" cap. "There are enemies committed to the demise of nuclear power," she warned them. In November 1986, more than 2,000 people marched over the bridge connecting Kennewick to Pasco bearing such signs as "Nuclear Power, Man's Best Friend" and "The Nuclear Industry is Safer than Farming and Logging." Their ire extended as well to organizations in the Tri-Cities and elsewhere that were arguing for nuclear disarmament. At a rally World Citizens for Peace held that August in front of the federal building, counterprotesters hung a banner proclaiming "How About Pearl Harbor?" on the opposite side of John Dam Plaza.

The link between Hanford and health effects has never been as clear-cut as many books and magazine articles have made it out to be. For one thing, many farmers and residents of the Tri-Cities knew that radiation was far from the only health hazard they faced. A familiar and entertaining sight in eastern Washington in those days was a single-engine plane swooping a few feet above a field of deep green crops to lay down a thin mist of pesticides. In the winter, when temperature inversions trapped pollutants close to the ground, the air turned brown and acrid from fertilizer plant emissions. Most farmers around the Tri-Cities got their irrigation water from the same place Hanford got its electricity—Grand Coulee Dam a hundred miles to the north. But irrigating the desert brought with it a plague of mosquitoes. To tamp down the infestation, trucks used to roll through the streets of eastern Washington towns emitting thick billows of DDT. Kathleen's parents were among the few who did not let their children ride their bicycles behind the foggers and breathe deep the clouds of sickly sweet pesticides.

The idea that Hanford was making people sick ran up against another problem. The Tri-Cities were full of people who were healthy, active, and playing golf well into their 80s. A common rejoinder to a comment about Hanford's health effects was: "Look at me, I've lived here my whole life and I'm fine." Careful study of Hanford workers did not find elevated rates of disease. On the contrary, because Hanford tended to hire healthy young people and pay them well, they suffered from fewer illnesses than did the average American. Long-range studies going all the way back to the Manhattan Project similarly could not find elevated cancer rates, nor could studies of populations living nearby. Exposure to Hanford's radiation might have affected some people, but the plant was not causing an epidemic of radiation-induced illnesses.

In 1986 the controversy over possible health effects blew up. That February, the Department of Energy released 19,000 pages of documents describing the history of Hanford's operations—part of a campaign by the department to quell continuing complaints about the plant. Government scientists and archivists had reviewed the papers to make sure that they would not reveal any secrets or cast Hanford in a bad light. Still, the documents showed that Hanford had released far more radioactivity into the air, water, and soil than outsiders had known. An example that made headlines across the Pacific Northwest was the Green Run. In December 1949, three months after the Soviet Union exploded its first atomic bomb, Hanford operators at the T Plant dissolved a ton of irradiated fuel elements just 14 days after they left the reactors. The radioactive iodine and xenon from the fuel elements spewed into the atmosphere, with planes and ground-based observers tracking the radioactive cloud. The idea was to figure out how much plutonium the Soviet Union was processing, under the assumption that the Soviet equivalent of Hanford had been dissolving green fuel to produce a bomb as soon as possible. But it rained on the day of the Green Run, and the wind shifted erratically, making the experiment hard to interpret. Meanwhile, radioactive iodine fell on plants, livestock, and people throughout eastern Washington State.

In 1987 the Department of Energy released a second batch of historical documents, just about the time it was becoming obvious that the N Reactor would never operate again. Newspaper and television reports were once more full of stories about Hanford's radioactive releases. A few years later, the first lawsuits were filed seeking damages for people who said they had been harmed by Hanford's emissions. Thousands more followed.

IN 1987, AT THE AGE of 58, Carolyn's father, Tom Deen, started feeling dizzy and tired. A local doctor diagnosed anemia, but when the usual treatment failed, he went to see a doctor at the University of Washington in Seattle. In early 1988, he received a new diagnosis: myelodysplastic syndrome, a cancer in which blood cells do not mature properly. Only about one person in 25,000 contracts the syndrome, usually around age 70, and one of the risk factors is exposure to radiation. He began receiving transfusions of blood, but he was often too weak to leave the house. He tried an experimental treatment, but it was painful, had uncomfortable side effects, and did not seem to make a difference. On November 1, 1988, at the age of 59, with his wife, daughter, and two sons at his side, he died. He rarely said anything negative about the industry in which he had worked his whole adult life. But just before his death he told Carolyn that maybe he had trusted the wrong people.

Carolyn's lifelong friend Kathleen had gotten married by this time, to Steve Flenniken, and together they had moved to the wet, western side of Washington State. As they began raising a family, Kathleen took a writing class, fell in love with poetry, and began writing poems. Later she wrote a poem about Carolyn's father:

To Carolyn's Father
—Thomas Jerry Deen, 1929–1988
On the morning I got plucked out of third grade
by Principal Wellman because I'd written on command
an impassioned letter for the life of our nuclear plants
that the government threatened to shut down
and I put on my rabbit-trimmed green plaid coat
because it was cold and I'd be on the televised news
overseeing delivery of several hundred pounds of mail
onto an airplane bound for Washington DC addressed
to President Nixon who obviously didn't care about your job

at the same time inside your marrow
blood cells began to err one moment efficient the next
a few gone wrong stunned by exposure to radiation
as you milled uranium into slugs or swabbed down
train cars or reported to B reactor for a quick run-in-
run-out and by that morning Mr. Deen
the poisoning of your blood had already begun.

Chapter 23

REMEMBERING

DEL BALLARD WAS OUTRAGED. THE DEPARTMENT OF ENERGY WAS
proposing to tear down the B Reactor. Didn't they know what that
facility represented? Tearing down the B Reactor would be like tearing down the Great Wall of China, or Stonehenge, or the Taj Mahal.
It was an irreplaceable artifact, a dividing line in history, an object
that explained how today's world came to be.

Admittedly, the Department of Energy had its hands full. In 1989,
it had signed an agreement with the State of Washington and the
US Environmental Protection Agency, known informally as the Tri-
Party Agreement, that required the department to clean up Hanford.
By then, plutonium production at Hanford was over forever. But
the site still had thousands of aboveground structures and buried
waste dumps containing radioactive materials, along with its tanks
of high-level wastes and contaminated soils. The department was
estimating that the cleanup would take 30 years and cost $50 billion.
From today's perspective, those estimates are laughably naïve.

Department of Energy managers had hard decisions to make. One
was what to do with the nine shut-down reactors that lined the right
bank of the Columbia River (technically, the N Reactor was still on
standby, but it was permanently closed in 1991). The department
could hire contractors to tear down the reactors and bury the pieces
in Hanford's desert soils. It could seal up the reactors to let their
radioactivity decay and demolish them sometime in the future. Or
it could bury them in place, creating immense concrete sarcophagi

scattered across the desert as permanent memorials to the dawn of the atomic age. But, as an early environmental impact statement from the Department of Energy put it, "no future long-term use of any of the [reactors] has been identified by the DOE." In other words, they were all slated for destruction.

In the summer of 1991, Ballard joined with a group of other Hanford supporters—including Jim Stoffels, who was as dedicated to preserving the B Reactor as he was to nuclear disarmament—to officially form the B Reactor Museum Association, or BRMA. At first the group's goal was simply to demonstrate that the reactor was a historic structure and should be preserved, and the next year they succeeded in placing it on the National Register of Historic Places. But that could not protect it from the Department of Energy's wrecking ball. They needed to get politicians involved.

At that time, the representative of the district that included the Tri-Cities was Sid Morrison, the scion of a fruit-growing family from nearby Zillah who served first in the Washington State legislature and then in the US Congress. Morrison was interested in making the reactor into a museum. He had fought hard over the past decade to keep federal funds flowing to Hanford. A preserved B Reactor could remind people of everything that had happened there.

Officials from the Department of Energy were resolutely not interested. "We're not in the museum business," they repeatedly told the members of the B Reactor Museum Association. Still, the department did not actively undermine the association's goals. In 1992, it decided to "cocoon" the reactors by enclosing them in concrete walls and metal roofs for up to 75 years, after which they could be torn down and buried. But the department always placed the B Reactor at the end of the cocooning list to give Ballard and his colleagues time to maneuver. "No one's heart at DOE was in tearing down the B Reactor," says Colleen French, a DOE manager who has long championed the reactor's preservation.

The members of the B Reactor Museum Association knew they needed some other organization to partner with the Department of

Energy if the reactor was to be opened to the public. Some thought a local museum could do it, but Ballard disagreed. The two organizations were too dissimilar. Only one institution could match the Department of Energy's clout, he thought—the National Park Service.

The idea seemed like a nonstarter. The Park Service had never partnered with the Department of Energy on a preservation project. Though it had experience establishing parks on controversial subjects—like the incarceration of Japanese-Americans during World War II at the Manzanar War Relocation Center in northern California—the Manhattan Project would be by far its most controversial undertaking. Just how controversial was suggested by the experiences of another federal agency. In the early 1990s, for the fiftieth anniversary of the end of World War II, the Smithsonian Institution made plans to display the *Enola Gay* at the Air and Space Museum in Washington, DC, with an accompanying exhibit that would explore the decision to drop atomic bombs on Japan and display artifacts from the bombed cities. The resulting fracas was the worst in the Smithsonian's history. Attacked by veterans' groups, politicians, and the media, the proposed exhibit soon collapsed. In the end, the display included only the fuselage of the *Enola Gay*, some videotaped interviews with members of the crew, and a history of the B-29 fleet.

The members of the B Reactor Museum Association knew that displaying the reactor would stir strong emotions. It was a symbol not just of technological triumph but of the threat of nuclear annihilation. They nevertheless persevered, despite the *Enola Gay* debacle. By the mid-1990s, Morrison had given up his seat in Congress to run unsuccessfully for governor of Washington State. The new congressman representing the Tri-Cities was Doc Hastings, who had owned and run a paper-supply company in Pasco before entering politics. Hastings was a master of the legislative process. In 2003, he managed to get legislation passed that directed the Park Service to study the feasibility of protecting Manhattan Project sites. Astoundingly, the draft study recommended that only the Los Alamos laboratory be made a fully managed park. Hastings went back to work. By the time the document was revised, the proposed park included all three of the major Manhattan Project sites—Los Alamos, Oak Ridge, and Hanford.

Such ideas take a long time to gestate and have many midwives. The Atomic Heritage Foundation, a nonprofit organization in Washington, DC, dedicated to preserving the history of the Manhattan Project and its legacy, made the creation of a national park its primary goal. Washington State's two senators, Patty Murray and Maria Cantwell, worked within Congress to advance the idea, with Cantwell's office drafting the initial legislation. The renowned historian of the nuclear age Richard Rhodes gave the keynote address at the sixtieth anniversary commemoration of the B Reactor. "You have, here in your midst, one of the world's most significant historical sites, a place where work was done that changed the human world forever," he told his enthusiastic Tri-Cities audience. Meanwhile, the Department of Energy cleaned up the B Reactor so that it was safe to enter, and the B Reactor Museum Association began taking people on public tours of the facility.

In 2013, unsuccessfully, and then again in 2014, Hastings, with support from the Tennessee, New Mexico, and Washington congressional delegations, introduced a bill to create a Manhattan Project National Historical Park. Hastings was retiring from the House in 2014, which gave the bill a momentum it would not have otherwise, and he chaired the House Natural Resources Committee, which had jurisdiction over the national parks. Included in the National Defense Authorization Act, the provision to establish the park passed on December 12, and President Obama signed it into law on December 19. The next year, the Department of Energy and the National Park Service signed a memorandum of agreement for joint management of the new park. The Energy Department continues to own and operate the three sites, but it agreed to facilitate public access to facilities in the park. The National Park Service assumed the job of interpreting the story of the Manhattan Project for the public and providing visitor services at the three locations.

Many questions continue to surround the park. Whose stories will it emphasize—those of the triumphant scientists and engineers who designed the bombs, those of the people who worked on the Manhattan Project, or those of the people against whom the bombs were used? Will it tell the stories of the people living near Oak Ridge, Los Alamos, and Hanford who claimed they were harmed by emissions from the facilities? What will it say about the radioactive wastes left behind and the many billions of taxpayer dollars that are still required to clean them up?

THE YEAR AFTER the Manhattan Project National Historical Park was created, the last of the lawsuits filed by people claiming that Hanford had harmed their health were finally resolved. Few of the downwinders were satisfied with the results. Some got cash settlements, but usually for less than they expected or needed. Others had their lawsuits dismissed or died before their cases were decided. The law firms that the government hired to defend itself made tens of millions of dollars—substantially more than the plaintiffs received.

The plaintiffs had always faced a difficult task. They had to prove that they would not have gotten sick if they had not been exposed to Hanford's radiation—in other words, that their exposure to excess radiation caused their cancers. But a cancer caused by radiation does not look different from a cancer caused by a carcinogenic chemical, by a genetic predisposition to cancer, or by bad luck. Some cancers are more common among people who have been exposed to radiation, but they occur as well in people who have not been exposed to radiation. Furthermore, everyone is exposed to radiation all the time—from the food we eat, the water we drink, the air we breathe, the objects around us, the medical procedures we undergo, even from outer space. To prove that excess radiation from Hanford caused particular illnesses required some sort of scientific evidence. That evidence could come partly from animal studies, but other animals are not perfect models for humans. No, the bulk of the evidence would have to come from epidemiology—the study of the health of large groups of people exposed to toxic or infectious agents.

Even before the first lawsuits were filed, just such a study was under way. In 1986, after the Department of Energy's release of documents describing Hanford's history, the federal government's Centers for Disease Control convened a group of researchers to talk about Hanford's possible effects on health. That review led to an $18 million initiative to determine, once and for all, if Hanford's radiation was responsible for some of the cancers and other illnesses of people living nearby. The researchers who led the Hanford Thyroid Disease Study, as it was called, decided to concentrate on the disease that should be easiest to find among downwinders. From 1944 through 1948, when filters were installed, and in lesser amounts thereafter, radioactive iodine-131—the radioisotope discovered by Glenn Seaborg and John Livingood back in 1938—billowed from the stacks next to the canyon plants whenever irradiated fuel elements were dissolved. Borne downwind, this iodine fell on vegetation, fruits, and vegetables. Cows and goats grazed on the vegetation, and the iodine passed into their milk. People in the area drank the milk and ate local fruits and vegetables. The iodine-131 concentrated in their thyroids, the butterfly-shaped gland in the neck that uses iodine-containing hormones to regulate growth and metabolism. There the radioiodine could irradiate thyroid cells and cause conditions ranging from benign nodules to cancer. Harmful effects are especially notable in children, since they drink more milk than other people and their thyroids are still growing. The researchers therefore decided to focus on people who grew up downwind from Hanford during the years of highest iodine releases.

Working with birth certificates from the area, the researchers identified more than 3,400 people who were born to mothers in seven Washington State counties between 1940 and 1946. They then gathered as much information as they could about the sources of the milk, water, and foods those people consumed. Reconstructing what someone ate and drank decades before is not easy, but the researchers were confident that they had estimated the doses within a factor of two or three. They then gave the study participants a thorough diagnostic evaluation for thyroid disease. By this time, other researchers had found higher rates of thyroid disease in residents of Hiroshima and Nagasaki exposed to the atomic bombs, in children downwind from Chernobyl, and in people who lived downwind from the Nevada Test Site. Leaders of the Hanford Thyroid Disease Study expected to find a similar effect in the population they were studying.

In 1999, they released a draft of their report to the public. Despite their best efforts, they could find no relationship between the doses of radioactive iodine people received and thyroid disease. In other words, people who were exposed to more radiation, according to their reconstructed doses, were not more likely to have a diseased thyroid. The researchers did find plenty of people with thyroid disease, but they concluded that the rates were not markedly higher around Hanford than in other areas.

The researchers who carried out the Hanford Thyroid Disease Study thought that their results would be welcomed in the Tri-Cities. As the initial draft of their report put it, the results should provide "a substantial degree of reassurance to the population exposed to Hanford radiation that the exposures are not likely to have affected their thyroid or parathyroid health." Instead, downwinders exploded in anger and frustration. "These 'reassurances' were worthless, even insulting, to the memory of loved ones dead of thyroid cancer or suffering with thyroid and parathyroid disease," wrote Trisha Pritikin, a resident of Richland in the 1950s whose immediate family had been ravaged by thyroid diseases. Making such a statement, she added, "was at best an exercise of very poor judgment and, at worst, just plain callous." Reviews of the study and subsequent research found slightly more thyroid disease in the area than in other areas, though few populations have been as rigorously screened for thyroid disease as the Hanford downwinders. Later studies also called into question whether the dose reconstructions were accurate, which would increase the findings' uncertainties. Still, after all the reviews of the initial draft were completed, the study's final conclusion stood: "These findings do not definitively rule out the possibility that Hanford radiation exposures are associated with an increase in one or more of the outcomes under investigation. However, it does mean that if such associations exist, they were likely too small to detect using the best epidemiologic methods available."

Radiation released by Hanford has certainly caused individual tragedies. Some people undoubtedly got sick and died because of excessive radiation they got from Hanford. Native Americans who lived in the area and ate lots of fish from the Columbia may have gotten especially high exposures. Some studies of nuclear workers and other people who are regularly exposed to excess radiation have found somewhat elevated levels of cancer and other diseases. Beyond the study's antiseptic language are harrowing individual stories of disease, disability, and death.

But the failure of the Hanford Thyroid Disease Study to find obvious signs of radiation-induced illness is, in retrospect, not surprising. Radiation in high doses, such as from the explosion of an atomic bomb, makes people sick and causes cancers. But radiation in low doses, such as the doses people received from working at Hanford or living nearby, is not a strong carcinogen. In the United States, about 42 percent of people will be diagnosed with cancer at some point in their lives, and about 20 percent will die from cancer. In a hypothetical case where all Americans received a one-time but fairly substantial dose of radiation—40 times the average background dose—about 43 percent of Americans would get cancer, according to the most widely accepted interpretation of past epidemiological studies. Furthermore, identifying the additional cancers caused by that sizable dose of radiation would be very difficult. People are exposed to so many carcinogens in their daily lives that separating the signal of cancers induced by low-dose radiation from the noise of other environmentally induced cancers is virtually impossible.

Fairness requires considering another possibility. Radiation harms a cell by ripping electrons off atoms, which can incapacitate or scramble the functions of a cell and possibly cause it to grow out of control. But cells have evolved biological mechanisms to repair such damage. In addition, cells exist within networks of other cells that can control aberrant cellular behaviors. At low levels of radiation, these repair mechanisms may be able to keep up with the damage caused by radiation and minimize or eliminate health effects. Epidemiologists may have been unable to find markedly elevated levels of disease around Hanford because the radiation released by Hanford has not caused markedly elevated levels of disease.

Most scientists who study radiation and health do not fear exposures to small doses of radiation, even if they recommend against unnecessary exposures. Hanford workers were mostly the same way. They knew they worked in radioactive environments. Most of the exposures they received were small, though sometimes they took risks to get something done. Most of them considered radiation part of their jobs.

Those observations accord with the experiences of people who worked on the Manhattan Project. Except for Enrico Fermi, most of the United States' atomic pioneers did not die prematurely from diseases that might have been caused by radiation. Glenn Seaborg died in 1998, at the age of 86, after suffering a massive stroke while exercising by walking up and down the stairs of the hotel where he was staying. Recognizing his condition as hopeless, he willed himself not to eat and died at home. A few years before his death, his lifelong quest to discover new elements received the ultimate accolade. The element with 106 protons, which he and his colleagues had detected in 1974, was named seaborgium. "This is the greatest honor ever bestowed upon me—even better, I think, than winning the Nobel Prize," he said. Yet for all the acclaim he received during his life, Seaborg will always be remembered as the discoverer of plutonium.

"Do I wish I hadn't discovered plutonium?" he once said. "No way. Once God had made a world that made bombs possible, there was no option. Both sides were going to make them. But if you ask me, 'Do I wish the laws of nature were such that you couldn't make an atomic bomb?' God, yes."

After a productive and peripatetic life that took her from the University of Chicago to the Institute for Advanced Study in Princeton to Brookhaven National Laboratory on Long Island to the University of Colorado to UCLA, Leona Woods died in 1986 at the age of 67 from a stroke. By that time, she had published more than 200 scientific papers on such varied topics as nuclear engineering, cosmology, and climate change. Her friend Herb Anderson, who had invited her to join the Met Lab in the summer of 1942, died two years later, on the forty-third anniversary of the Trinity test, at age 74. He had suffered for decades from the lung disease berylliosis, which he contracted from breathing in beryllium while making neutron sources with Fermi.

Leo Szilard died at 66 of a heart attack, though he also had developed bladder cancer by then. His nemesis Leslie Groves died six years later at 73 from heart problems that had plagued him for years.

Robert Oppenheimer died at 62 from throat cancer, almost certainly caused by his incessant cigarette smoking. Ernest Lawrence died even younger, at age 57, from a combination of colitis and atherosclerosis; element 103 is named lawrencium in his honor. Arthur Compton died at 69 from a cerebral hemorrhage after a distinguished career as chancellor and professor at Washington University in St. Louis. And Raisuke Shirabe, who was standing a half mile from the bomb that detonated over Nagasaki and nearly succumbed to acute radiation poisoning, died in 1989 shortly before his ninetieth birthday.

Rumors about secret government projects had a tendency to run wild even before the internet was invented, and one was that all the people who witnessed the startup of Chicago Pile 1 beneath the west stands of Stagg Field died early of cancer. On the contrary, more than two-thirds of the 49 people who were there that day lived for at least 36 more years, and more than half lived more than 50 years longer. Of those for whom the cause of death is known, only seven died from cancer, and their cancers were probably not caused by radiation. At least 13 of the people who were with Enrico Fermi that freezing cold day in Chicago died in their nineties.

After the death of Tom Deen from myelodysplastic syndrome, Carolyn and the rest of her family decided not to file a lawsuit seeking damages. "I didn't want their blood money," Carolyn said. After Carolyn's mother died, she moved back into the family house, on Cedar Street in Richland, while continuing to work at Columbia Basin Community College as a trainer of elementary school teachers.

A few years later, a close relative of Carolyn's who had worked at Hanford contracted a cancer sometimes caused by exposure to radioactivity and other toxins. So far, treatment appears to have stemmed the progression of his disease.

●

IN THE 1980S, when she was 38 years old, Susan Leckband moved from Iowa to Richland to go to work for a small long-distance telephone company. She immediately fell in love with the stark natural beauty of her new home: the symphonic skies, awash in color and light; the nearby hills and more distant mountains, brown most of the year but a deep mossy green after the rains of winter; the magic blue ribbon of the Columbia River, a crease in the desert that has been attracting people to the area for millennia. A few years later, Leckband applied for a job at Hanford. She did not have a college degree, but she soon was hired and began acquiring jobs of increasing responsibility. She did administrative work involving the Plutonium Finishing Plant in the 200 area, the PUREX separation facility, the water-filled basins that held the spent fuel from the K Reactors. Supervisors recognized her competence and rewarded her for it.

In 1993, Hanford held a celebration to mark the fiftieth anniversary of the plant, and Leckband volunteered to help re-create a mess-hall dinner. The auditorium at the Benton County fairgrounds was decorated with World War II posters, and servers ran platters piled high with food to tables of old men and women who had arrived at Hanford a half-century earlier. "I was so moved by their patriotism, their love of their country," Leckband recalled. "They moved away from comfortable homes to live in conditions that most of us would consider worse than camping."

She got interested in an unusual institution that marked the transition from plutonium production to cleanup at Hanford. The Tri-Party Agreement signed by the Department of Energy, the Environmental Protection Agency, and the Washington State Department of Ecology led in 1995 to the creation of a Hanford Advisory Board that would produce guidance for each of the three agencies. An intriguing experiment in civic engagement, the board contains seats for people representing local and state governments, business interests, the Hanford workforce, environmental organizations, public health agencies, tribal governments, universities, and the public at large. Despite its diversity of interests, the board operates by consensus. On the rare occasions where it cannot reach agreement, it conducts what it calls a sounding board to gather and make public the views of its members.

Leckband applied for the board position designated for a "non-union, non-management" member and was selected in 1997. Since 2002, she has been either vice chair or chair of the board. She runs tight meetings with strict rules of engagement. Her number-one rule is that people will speak civilly to each other, even when they disagree. "People are passionate. I respect diverse opinions. But it's my job to get the outcome we really want, which is consensus."

Since its establishment, the Hanford Advisory Board has issued more than 300 consensus statements on everything from the leaking waste tanks to budget priorities for the cleanup to worker safety to the long-term stewardship of Hanford's land. "Sometimes it's painful," Leckband said. "Having someone say, 'Your baby is ugly,' that can cause friction. Some officials are not comfortable with the public pushing back, because their boss is pushing on them. We on the board need to respect both sides. But our job is to say that this is what the public needs."

In recent years the board's greatest concern has been the amount of work that remains to clean up Hanford, despite the money the federal government and taxpayers have already devoted to the task. The leaders of the Manhattan Project did not devote much thought to the mess they were creating. They had a war to win; other people could worry about the environment later. Later, the companies that operated Hanford during the Cold War were more concerned with producing plutonium than disposing of wastes properly. Now the bill has come due—and it is immense. The Department of Energy has calculated that cleaning up Hanford will cost at least $300 billion and possibly more than $600 billion, much more than the cost of building and operating Hanford throughout its history. Today the federal government is spending about $2.5 billion a year to clean up Hanford. In one of its guidance documents, the Hanford Advisory Board observed that funding will need to increase to more than $9 billion a year to get the job done. This seems unlikely given the federal government's many other obligations.

The government has made significant progress on the cleanup, despite how far it still has to go. Six of the nine reactors have been cocooned, and two more will be soon. Contractors have dug up old dumps and contaminated soils and have moved the wastes to a huge landfill near the separation plants, where the radioactive byproducts of Hanford's history can be isolated and monitored. Gigantic pump-and-treat stations scattered across the site lift water from below ground, remove contaminants that can go into landfills, and reinject the clean water back into the subsurface. Except for a few troublesome plumes and contaminants, the treatment plants have largely stemmed the flow of radioactive and chemical toxins toward the Columbia.

But the hardest and most expensive cleanup tasks are just getting started. In particular, the high-level radioactive wastes generated during the Manhattan Project and Cold War continue to sit in the single-shelled and double-shelled tanks around the canyon buildings. The plan has been to mix the waste with glass flakes, heat the mixture, and send the resulting glass logs to a high-level waste repository. But this vitrification technology has been expensive and hard to develop, and the Department of Energy and its contractors have spent billions of dollars on botched attempts to get vitrification to work. Furthermore, the vitrified waste has no place to go, since the United States has not yet designated a place to dispose of its high-level military and civilian nuclear wastes. Even after the waste is vitrified, it will need to be stored on site until a repository is available.

Most recently, the Department of Energy has proposed to vitrify most but not all of the wastes from the tanks and then reclassify the remaining hard-to-remove waste as low-level waste, after which it would fill the tanks with grout, put a fence around the site, and try to keep people away far into the future. Officials and advocates from Washington and Oregon, where hundreds of thousands of people live downriver from Hanford, have strenuously objected. Over the years, they point out, rainwater will penetrate the tanks, leach through the grout, and carry radionuclides toward the river. Leaving waste in the tanks would essentially create a shallow high-level nuclear waste repository at Hanford, saddling the region with a mess that the federal government has vowed, ever since World War II, to rectify.

Leckband wants what other people in the Pacific Northwest want—for the federal government to fulfill its promises. She wants her children and grandchildren to be able to hike and ride their bikes along the river, past the shuttered reactors and separation plants, without worrying about radioactivity in the water, soil, or wind-blown dust. She wants the local tribes, several of which are represented on the Hanford Advisory Board, to regain access to the cultural sites and traditional hunting and fishing grounds that they have been guaranteed in treaties. Several of the facilities left at Hanford remain extremely dangerous. A collapsed waste tank, a fire, an earthquake—many kinds of disasters could spew radioactivity across the landscape. Preventing such catastrophes will require careful planning, empowered workers, and increased funding.

Ironically, the Hanford nuclear reservation is one of the best-preserved regions in all of eastern Washington. The areas associated with the reactors and separation plants take up just a small part of the overall site, which was half the size of Rhode Island when it was established. The rest of the land has gone largely untouched since Matthias flew over Richland on that first day of winter in 1942. A proposal to build a dam near the Tri-Cities was scuttled when people observed that the resulting reservoir would raise the water table beneath Hanford's wastes. As a result, the 50-mile stretch of the Columbia that runs through Hanford is the last free-flowing section of the river from the Canadian border almost to Portland, where Lewis and Clark noticed the tides of the Pacific beginning to lift their canoes. In 2000, President Bill Clinton preserved almost half of the reservation as the Hanford Reach National Monument. Today people can hike on the chalky white cliffs across the Columbia from the cocooned reactors and watch bald eagles soar above the wind-rippled river.

As late as the 1960s, the population of the Tri-Cities area was just 50,000 people. Today, more than 300,000 live there, and the region is booming. People from the soggy western side of the state have retired there, attracted by its sunshine and conservative politics. The economy is strong, partly because of the cleanup funds flowing to the region. Almost four thousand people work at Pacific Northwest National Laboratory just north of Richland on everything from ecosystem science to energy efficiency to nuclear nonproliferation. A few miles northwest of the lab, the Laser Interferometer Gravitational-Wave Observatory searches the cosmos for signals from colliding black holes and neutron stars. Perhaps most surprising to people who have lived in the region for decades, the barren hillsides around the Tri-Cities, which are at about the same latitude as central France, have proven ideal for growing grapes. Today, on weekends, the area is packed with people who have come to the Tri-Cities to sample some of the world's best wines.

◉

A FEW DAYS BEFORE the atomic bombing of Nagasaki, Mitsugi Moriguchi's mother decided that she needed to get her children as far away from the city as possible. The Mitsubishi shipyards in Nagasaki had become a target of conventional US bombers, and the family lived nearby. In 1945, Mitsugi was eight years old. His oldest brother was in the military, another brother and a sister were middle-schoolers and therefore required to work in munitions factories, an older sister was in the sixth grade, and a younger brother was in the first grade. Many of the families near them were evacuating to the northern part of the Urakami Valley, but Mitsugi's mother did not think that was far enough away. She decided, instead, to take her three youngest children into the countryside—she knew that she would have to leave her middle-schoolers behind, since they could not be excused from their factory jobs. Three days before the bombing, Mitsugi, his two siblings, and his mother got onto a train. When they got off they were 25 miles away from the city.

When the bomb detonated over Nagasaki, they heard the explosion and saw an enormous cloud rising over the city. A voice on the radio said, "Citizens of Nagasaki, get out, get out." Then the radio went silent. Immediately, Mitsugi's mother made plans to return to the city to find the husband, son, and daughter she had left behind. The children pleaded with her to stay. They could not survive without her, they cried. But she said that she had to go. Before she left, she gave them all the money she had and said, "If none of us survive, you must use this cash to do the best you can."

For three days, Mitsugi and his older sister and younger brother waited at the train station for their mother to return. A station master told them, "Nagasaki has been completely destroyed." They talked among themselves about what to do. They were about to give up but did not know where to go. On their third night at the station, a hand shook Mitsugi awake in the middle of the night. He looked up and saw a ghost—a woman covered with dust, whose clothing was in shreds, her hair wild and flying about her face. It was his mother. Behind her stood his older brother and sister. She had found them and brought them back.

They lived by the train station for several days. Mitsugi's brother had been severely burned when the bomb destroyed the munitions factory where he was working. They had no medicine, so they applied grass to his wounds. His sister had been working in a wooden building that collapsed, but she was able to crawl out from under the debris.

Someone told them that the war was over, and they decided to return home. As they walked through the cleared streets of Nagasaki, Mitsugi saw that nothing was left. They walked past demolished structures, through the smoke from funeral pyres. Mitsugi remembered the smell of burning flesh for the rest of his life. When they got to their house, they saw that the roof was gone but that the walls were still standing. Their father had survived. He had stayed home from his factory job the day of the bombing because the roof needed repairs.

Seventy-three years later, retired from his lifelong job as an elementary school teacher, Moriguchi came to the Tri-Cities to tour the B Reactor and talk about the atomic bombing of Nagasaki. The trip was organized by a group called Consequences of Radiation Exposure, which Pritikin had established to work on issues important to Hanford downwinders. The B Reactor is not easy to reach—it requires a 30-minute bus ride each way from the edge of Richland—but it is by far the most impressive site in the Manhattan Project National Historical Park. It sits a half-mile or so from the Columbia River, its cinder block walls the color of the surrounding desert, a cenotaph to the horrors of war. If you know what you're looking at, walking through the front entrance into the large room facing the reactor, where operators used to load uranium fuel elements into the process tubes, can take your breath away. The reactor is almost exactly the same as it was the day Enrico Fermi, Leona Woods, and Crawford Greenewalt started it up and watched it slowly succumb to xenon poisoning. The B Reactor Museum Association has placed exhibits in the building to explain its significance, which the National Park Service eventually will supplement and replace. As would be expected of displays erected by a group of engineers, they focus on the construction and operation of the machine, the technological feats it accomplished, and the ingenuity it embodies. The displays do not mention the destruction of Nagasaki or the continuing threat of nuclear annihilation posed by bombs containing Hanford's plutonium.

One of Moriguchi's hosts at the B Reactor the day he visited was 90-year-old John Fox, a former mayor of Richland, a member of the B Reactor Museum Association, and an engineer who worked at Hanford for decades. At the urging of the Consequences of Radiation Exposure group, Moriguchi had brought a handheld dosimeter with which to measure the building's radiation. He told Fox that if he stayed in the building for a year, his exposure would be greater than safety standards for the general public permitted. "We won't keep you here for a year," Fox replied.

Speaking through a translator, Moriguchi seemed to be alternately amused and appalled by the reactor. He sat in the chair in front of the reactor control panel and had his picture taken. He wondered why the reactor displays made no mention of the suffering caused by the bombing of Nagasaki. "We learned it was going to become a national park, and we in Nagasaki are quite worried," he said. "Was it going to become a national park to express pride? Or to promote reflection?"

Fox and Moriguchi talked as they made their way through the reactor building. Fox told him about his work removing the fuel elements from the reactor and sending them to the canyon buildings. Moriguchi spoke about his sisters' miscarriages in the years after the war and about the cancers that had ended his siblings' lives. In 1945, Fox said, he was a young man about to enter the military and be sent to the Pacific. The bombing of Nagasaki "saved me from being drafted and participating in an invasion of Japan and ending up there dead on a beach," he said.

"I have nothing against patriotism," said Moriguchi. "But I want people, in addition to loving their country, to love human beings, to love humanity."

The two men were not angry or accusatory. Their voices were wistful, as if their thoughts were far away. They were not asking for apologies. They seemed, rather, to be seeking simply an acknowledgment of what had happened, of what they had experienced, of what this place meant. At the end of the tour, both men had tears in their eyes. They stepped forward and hugged each other.

EPILOGUE

ENRICO FERMI WAS A MAN WHO LOVED TO SOLVE PROBLEMS. ONE DAY in 1950 he was walking with three other physicists to lunch at the Los Alamos laboratory. Inspired by a recent cartoon about flying saucers in *The New Yorker*, they were discussing whether extraterrestrial beings could travel from one planetary system to another. By the time they sat down for lunch, the conversation had turned to other subjects. Then, out of nowhere, Fermi asked, "Where is everybody?"

The others all laughed, because they knew what Fermi had done. He had run the numbers in his head. He had been thinking about the immense number of stars in the galaxy and the galaxy's great age. Say that an advanced civilization evolved on a planet around one of those stars, and that this civilization developed the ability to travel from its own planetary system to another. Maybe it would take a spaceship a few hundred years to reach another star, maybe a few thousand. But once members of this civilization reached another planetary system, they (or more likely their descendants) could use it as a steppingstone to keep going. Certainly within a million years or so, Fermi figured, a technologically advanced civilization could spread to every habitable planet in the galaxy.

But our galaxy is thousands of times older than that. Therefore, if such a scenario has even a slight chance of happening, it almost certainly would have happened by now. Since we have no good evidence of extraterrestrial visitors, the spread of intelligent beings across the galaxy must never have happened. This argument—that extraterrestrial visitors should have visited Earth but appear never to have done so—has become known as the Fermi Paradox.

People have proposed dozens of solutions to the Fermi Paradox. Maybe we are the only technologically advanced civilization that has evolved in the galaxy. Interstellar travel could be too difficult or expensive for civilizations to undertake. Maybe intelligent creatures quickly realize that broadcasting their existence to the rest of the galaxy is not a good idea, or maybe we've been quarantined from the galactic community for some reason.

But the grimmest explanation seems the most likely. Advanced civilizations may not last long enough to launch themselves into space. Maybe they deplete their planet's resources and decline. They might succumb to volcanic eruptions, meteoroid impacts, or exploding stars. Or, as soon as they discover nuclear energy, they blow themselves up.

We have already come close to destroying ourselves several times. Reflecting on the Cuban missile crisis, President Kennedy thought that the odds of worldwide nuclear war had been at least one in three. One premise of the Fermi Paradox is that even unlikely events will happen eventually. If future international crises have at least some possibility of going nuclear, human civilization will not last for long. People may be reassured that the number of warheads on Earth has declined by three-quarters since the end of the Cold War. But even before the bombings of Hiroshima and Nagasaki, the leaders of the Manhattan Project knew that numbers don't matter with atomic bombs. Even a limited exchange of nuclear weapons, besides killing millions of people, would loft enough smoke into the atmosphere to cause widespread famine. A full-scale nuclear war would end food production on the planet for years. If the United States and Russia fired even a fraction of their warheads at each other, human civilization would likely end. And if we really are the only civilization in the galaxy, such an event would have cosmic significance.

Given the possibility that any use of nuclear weapons could rapidly escalate into a more general conflagration, the only way to safeguard human civilization is to eliminate all nuclear weapons from the Earth. The best way to do that will be to rid the world of the materials bequeathed to it by the Manhattan Project—the uranium-235 and plutonium-239 that can be used to make atomic bombs. That does not necessarily mean abolishing nuclear power, which will almost certainly be needed to counter the potential calamity of global climate change. But it does mean eliminating the chemical process Glenn Seaborg and his colleagues pioneered in their Berkeley laboratory during that stormy night of February 24, 1941—the separation of plutonium from irradiated uranium. The reprocessing of reactor fuel to yield plutonium will always pose the risk that atomic bombs could be quickly reconstructed in a nuclear weapons–free world, whether by a breakout nation or by terrorists who somehow gain access to plutonium. Plutonium generated in nuclear reactors is safeguarded by the intense radioactivity of the fission products in fuel elements. But reprocessed plutonium that has been removed from fuel elements is too dangerous to permit its continued existence in this world.

The abolition of nuclear weapons and all remaining fissile materials will require something that has been the antithesis of US policy since Seaborg and his fellow scientists quit publishing their research in the scientific literature at the beginning of World War II. The United States and all other countries will need to open their nuclear facilities, both civilian and military, to international inspections. All countries will need to agree to a verifiable and enforceable treaty that bans the manufacture and possession of fissile materials and nuclear weapons. All nations will need to agree to intrusive and rigorous inspections of their nuclear facilities. Any nation that violates the treaty will need to be subject to economic sanctions and to military responses with conventional forces.

Many groups are working to achieve these goals. Direct action groups like Jim Stoffels's World Citizens for Peace continue to draw attention to the threat posed by nuclear weapons. Groups founded by scientists, physicians, and other professionals are working for the complete elimination of nuclear weapons and fissile materials. A decade of advocacy by the International Campaign to Abolish Nuclear Weapons led to the Treaty on the Prohibition of Nuclear Weapons, which was adopted by the United Nations in 2017. The nations that possess nuclear weapons have not signed the treaty, severely limiting its effectiveness. But it establishes a goal toward which the world can strive.

Yet the United States and Russia—which together possess more than 90 percent of the world's nuclear weapons—are headed in exactly the wrong direction. They have steadfastly refused to adopt nuclear abolition as an objective. On the contrary, they continue to miniaturize and modernize their nuclear weapons, making it ever more likely that the moral abomination of nuclear war will someday occur.

Beyond the threat that nuclear weapons pose to all humans and other living things is their outrageous cost. If even a portion of the trillions of dollars and rubles spent on such weapons had been devoted instead to health care, education, housing, transportation, energy research, and pollution control, people in the United States and in the former Soviet Union would lead better lives today, as would people around the world. Americans continue to spend an average of hundreds of dollars per person per year to maintain and upgrade our nuclear weapons. The decision to spend those resources on instruments of destruction has been an immense historical tragedy.

I entitled this book *The Apocalypse Factory* not to shame the leaders of the Manhattan Project or the proud and resourceful workers at Hanford. Rather, I'm using the word *apocalypse* in its original sense. In the Bible, the apocalypse is not the final battle between good and evil—that's Armageddon, a word derived from an ancient military stronghold on a trade route linking Egypt and the Middle East. An apocalypse is a revelation—literally an uncovering—about the future that is meant to provide hope in a time of uncertainty and fear. Apart from the brilliance of its technological achievements, the story of Hanford is mostly a story of human misunderstanding, belligerence, and short-sightedness. Yet nuclear annihilation has not yet occurred. We have learned how to generate electricity using nuclear energy, and future reactor designs promise major improvements in safety. Many nuclear sites have been cleaned up, even if the remaining cost of cleaning up Hanford is daunting. We have many more things to clean up in this world, not just radioactive wastes but chemical wastes, dangerous biological agents, and the hundreds of billions of tons of carbon dioxide that we have emitted into the atmosphere. Hanford's cleanup, if done persistently and well, could provide an object lesson in making the Earth whole again.

One time, Fermi was walking with a group of friends next to an irrigation canal near Hanford. Such canals are common in the region. They carry water from the Columbia River to croplands, often quickly to avoid evaporation. Fermi was a good swimmer. He wondered if he could get into and out of such a canal without being swept away. Before his companions could talk him out of it, he took off his shirt and jumped into the canal. He was swept away.

I've been in those canals. They're killing machines. They carry you along too swiftly to swim. The concrete sides of the canals are mossy and smooth. There's nothing to grab onto. Kids I knew drowned in those canals when the water swept them down a causeway or passed through a culvert.

Fermi's companions ran along the side of the canal, trying to catch up with him. Soon they saw him in the distance. He was standing on the road next to the canal, bleeding and shaken.

We can get out of the mess we've created. It won't be easy. It will take all the ingenuity we have. But we are ingenious creatures.

ACKNOWLEDGMENTS

In 1984, Allen Hammond, editor of the magazine *Science 84*, sent me to Hanford to report on the decommissioning of the Shippingport, Ohio, nuclear reactor, the first commercial reactor in US history. As a Department of Energy official and I were driving across the sageland to the trench in which the reactor would be buried, we began passing, in the distance, immense concrete buildings. "What are *those*?" I asked him.

"They're the plants where we made plutonium for our nuclear weapons," he told me.

That has to be a great story, I thought.

It took me a long time—and a move from Washington, DC, back to the state where I grew up—to figure out how to write about Hanford, but when I finally mentioned the idea to my friend and agent Rafe Sagalyn, he liked it immediately. So did my talented and insightful editor at W. W. Norton, Alane Mason. Over the ensuing months and years, Rafe and Alane both helped me shape and sharpen the story I had to tell. This book would not exist without their hard work.

Also at Norton, Mo Crist, Julia Druskin, Amy Medeiros, Laura Goldin, Beth Steidle, and Sarahmay Wilkinson shepherded the book through production. Gary Von Euer copyedited the manuscript, Yang Kim designed the cover, and Rachel Salzman handled publicity.

The Sloan Foundation, through its Public Understanding of Science, Technology, and Economics program, provided me with a grant to support the researching and writing of this book. I'm very grateful to Doron Weber and his colleagues for their assistance.

A book like this would not be possible without the help of librarians and archivists, who continually amaze me with their enthusiasm in responding to the questions of authors. At the University of California, Berkeley, I'm grateful to the staff of the Bancroft Library for providing me with access to the journals of Glenn Seaborg. Also at Berkeley I'd like to extend my thanks to Doris Kaeo and Esayas Kelkile for a guided tour of Room 307 in Gilman Hall, where Glenn Seaborg and his colleagues first isolated plutonium that stormy night of February 24, 1941.

At the Department of Energy's Public Reading Room in the library at Washington State University–Tri Cities, Janice Scarano and her staff—Teresa Hall and Doug Sharpe—spent many hours with me going through materials on Hanford that they and their predecessors have been compiling and safeguarding for decades. The librarians at the Richland Public Library, including Ann Roseberry and Gavin Lightfoot, were very helpful in providing me with access to the materials in the Richland Collection.

At the Library of Congress, Patrick Kerwin and his colleagues made it possible for me to work with the 1,020 boxes of materials that Glenn Seaborg left to the library. At the University of Chicago's Regenstein Library, Greg Fleming and Daniel Meyer were equally helpful guides to the papers of Herb Anderson, Sam Allison, and Enrico Fermi and to the other resources in the library's special collections. At the Hagley Museum and Library in Wilmington, Delaware, Lucas Clawson provided me with the papers of Crawford Greenewalt and Franklin Matthias and talked with me at length about DuPont's role in building and operating Hanford. In the archives office of the National Academy of Sciences, Janice Goldblum provided me with materials from the committee that studied the feasibility of atomic bombs in 1941.

At the libraries of the University of Washington and Washington State University, Trevor Bond, Anne Jenner, and many other librarians and archivists provided me with documents, pointed me toward things I needed, and otherwise made researching the history of Hanford a pleasure. Similarly, at the Seattle Public Library, Linda Johns and Ann Ferguson helped me find documents that are available nowhere else.

Writers in Seattle are lucky to have access to such superb libraries and accomplished librarians.

Telling the story of the atomic age from the perspective of Hanford and plutonium has required recounting, usually with a different emphasis, episodes that are familiar to experts on the Manhattan Project. For their previous scholarship I'm grateful to Kate Brown, John Findlay, Michele Gerber, Bruce Hevly, William Lanouette, Robert S. Norris, Cameron Reed, Richard Rhodes, J. Samuel Walker, Alex Wellerstein, G. Pascal Zachary, and especially Cynthia C. Kelly, the founder and president of the Atomic Heritage Foundation, whose websites continue to be by far the best single source of information on the Manhattan Project and the early years of the Cold War.

In Nagasaki, Susumu Shirabe, Hitomi Shirabe, and Mariko Mine provided invaluable assistance and documents as I retraced the steps of Hitomi's grandfather, Raisuke Shirabe, in the hours, days, and months after the bombing. Atsumi Nishimura provided me with Japanese translations throughout the writing of this book.

Many people who have worked at Hanford or have lived in the Tri-Cities shared their stories and expertise with me, including Terry and Jim André, Tom Bailie, Del Ballard, Randy Bradbury, Carolyn Fazzari, Kathleen Flenniken, John Fox, Roy Gephart, Harvey Gover, Ron Kathren, Greg Koller, Pamela Brown Larsen, Susan Leckband, John McCloy, Larry Morgan, Bruce Napier, Gary Petersen, Trisha Pritikin, Alan Rither, Mark Smith, Don Sorenson, Jim Stoffels, and Gene Weisskopf.

For their generous and expert assistance I especially want to thank Robert Franklin, Jillian Gardner-Andrews, and Michael Mays at the Hanford History Project, Becky Burghart and Kris Kirby at the National Park Service, Colleen French at the Department of Energy, Tom Carpenter and Liz Matson at Hanford Challenge, Steven Gilbert at the Institute of Neurotoxicology and Neurological Disorders, Jon Brock at the University of Washington, Robert S. Norris at the Federation of Atomic Scientists, Alex Wellerstein at the Stevens Institute of Technology, Cameron Reed at Alma College, Annette Cary at the *Tri-*

City Herald, Hal Bernton and Sandi Doughton at the *Seattle Times*, Joe Copeland, Jenny Cunningham, and Knute Berger at *Crosscut*, Michael Krepon at the Stimson Center, Ken Niles at the Oregon Department of Energy, and Eric and Ellen Seaborg.

My daughter, Sarah, and son, Eric, spent many hours helping me do research on the Manhattan Project, the Cold War, Hanford, and Nagasaki. Sarah and Seattle-based mapmaker Matt Stevenson drew the wonderful illustrations in the book.

Five people read and commented on this book when it was at a critical juncture: Robert Franklin, Sally James, Cynthia Kelly, John Martin, and Melanie Roberts. Their feedback had a big influence on this book. Many friends also offered much-needed advice and encouragement, including Joe Alper, Thomas Conkling, Gregg Easterbrook, Susan Feeney, Kevin Finneran, Donna Gerardi Riordan, Greg Graffin, David Jarmul, Marjorie Kittle, Jill Lawrence, Priscilla Long, Lisa Olson, Rick Olson, Dan Para, Blake and Connie Rodman, Eric Scigliano, Jack Shafer, Mark and Judith Stein, Umberto Vizcaino, and David Williams.

I have dedicated this book to my wife, Lynn, without whom none of my books would have been written. I have also dedicated it to the memory of John Hersey, from whom I took a course on narrative nonfiction in 1978. Even now, sitting at my desk, I can hear his calm and patient voice offering me advice.

NOTES

The following notes refer to books, articles, publicly available interviews, and other documents listed in the bibliography. Documents only available in archives are described in full in the notes. Quotations in the text that do not have a corresponding note are from interviews that I conducted.

One of the great pleasures of writing about historical events in the age of the internet is that many historical documents, and especially documents associated with the use of atomic bombs at the end of World War II, are available online. Rather than citing in the notes the physical locations of these documents, I have provided enough information in the text to find them with a quick web search. I invite readers to consult the full documents and draw their own conclusions about the world-changing events they describe.

PROLOGUE

1   As soon as Franklin Matthias flew over the Horse Heaven Hills: Sivula, "Fateful flight selects site," A1.

2   "This is it": Franklin Matthias, interview by B Reactor Museum Association (BRMA), September 26, 1992. Available upon request to BRMA.

2   Even as Matthias was flying over the towns of Richland, Hanford, and White Bluffs: Jones, *Manhattan*, 78–85.

4   "most contaminated nuclear site in the Western Hemisphere": Gilbert et al., *Particles on the Wall*, 3.

4   "I treated plutonium production": Rhodes, "Hanford and History."

5   "known sin": Oppenheimer, 88.

5   "My God, if everybody that has made an important contribution": Interview of Crawford Greenewalt by David A. Hounshell and John K. Smith,
December 15, 1982, Hagley Museum and Library, Accession 1878. Available upon request to the Hagley Museum and Library.

PART 1: THE ROAD TO HANFORD

7   "I didn't think, 'My God'": Seaborg, "Glenn T. Seaborg Biography
1912–1999."

CHAPTER 1: BEGINNINGS

9   Toward the end of his long: Hoffman, 247.

9   "pathways to success in big business": Hardy, 498.

9   "Seaborg's brother is a truck driver!": Segrè, interview by Richard Rhodes.

10   "We weren't poor": Seaborg and Seaborg, 10.

10   "taught chemistry with the charisma": Ibid., 13.

10   "You could learn certain principles": Seaborg, interview by the American
Academy of Achievement.

11   "lingering disbelief": Seaborg and Seaborg, 22.

11   "The whole Berkeley atmosphere": Yarris.

11   "Berkeley was Wonderland": Seaborg and Seaborg, 23.

12   The morning after his evening in the library: Hiltzik, 47.

13   "I got out of there just in time": *Time Magazine*, 72.

13   They were sometimes called the "nim nim boys": Bird and Sherwin, 94.

13   "I couldn't get over the feeling": Seaborg and Seaborg, 27.

15   In January 1934 the husband and wife team: Guerra et al., 49–51.

16   Over the next few years: Seaborg, "My Career as a Radioisotope Hunter,"
962–64.

17   "I wonder if any of the people": Jolly, 157.

CHAPTER 2: THE CHAIN REACTION

19   "I have something terribly important to tell you": Alvarez, 72.

19   "When Alvarez told me the news": Wilson, *All in Our Time*, 28–29.

19   "If you hit a car-size boulder with a pick": Seaborg and Seaborg, 56.

19   "I walked the streets of Berkeley": Ibid., 59.

20 "What an exciting specialty I'd chosen": Ibid., 60.

22 "How can anyone know what someone else might invent?" Lanouette, 133.

23 "there was very little doubt in my mind": Szilard, *His Version of the Facts*, 55.

24 "It would take the entire efforts": Frisch and Wheeler, 52.

26 "It seems to me now": Szilard, *The Collected Works*, 193.

26 The physicists at Columbia "started looking": Fermi, *Collected Papers, Volume II*, 1000.

CHAPTER 3: ELEMENT 94

29 Well past midnight on February 24, 1941: Seaborg, *Journals*, 28–29.

29 "a less significant or historical looking room": Seaborg, *Nuclear Milestones*, 3.

31 "can be separated from all the known elements": Seaborg, *Journals*, 29.

31 Seaborg later recalled: Bickel, 188.

31 "I was a 28-year-old kid": Seaborg, "Glenn T. Seaborg Biography 1912–1999."

32 On Monday, March 3, 1941: Seaborg, *Journals*, 30.

34 "Fortunately, we were spared": Seaborg and Seaborg, 72.

34 "I was unfamiliar with the god": Ibid.

CHAPTER 4: THE DECISION

36 he considered his first name "a nuisance": Zachary, *Endless Frontier*, 20.

36 "you couldn't get anything done in that damn town": Goldberg, "Inventing a Climate of Opinion," 451.

36 "this uranium headache": Zachary, *Endless Frontier*, 189.

37 "I with that the physicist who fished uranium": Ibid., 189.

37 "scared to death": Ibid., 195

37 "As long as I am convinced": Compton, *The Cosmos of Arthur Holly Compton*, 44.

38 "within twelve months": National Academy of Sciences, Committee on Atomic Fission, May 17, 1941, report. Available upon request to the National Academy of Sciences archives.

38 "the first side to perfect this scheme": Parides, 27.

39 "I still shudder when I think": Bush, 279.

39 "If large amounts of element 94 were available": National Academy of Sciences, Committee on Atomic Fission," July 11, 1941, report, "Memoran-

dum Regarding Fission of Element 94" by Ernest O. Lawrence. Available upon request to the National Academy of Sciences archives.

40   "I knew that the effort would be expensive": Bush, 59.

40   "of superlatively destructive power": National Academy of Sciences, Academy Committee on Uranium. Available upon request to the National Academy of Sciences archives.

41   "Seaborg tells me that within six months": Compton, *Atomic Quest*, 71.

CHAPTER 5: THE MET LAB

43   When Seaborg stepped off the train: Seaborg, *Journals*, 109.

43   Organized and led by Jimmy Doolittle: Scott, 38.

44   "a stack 200 to 300 feet high": Seaborg, *Journals*, 121.

45   "We're working on something that's more important": Seaborg, interview by Stephane Groueff.

45   "Some stare in disbelief": Seaborg, *Journals*, 155.

46   "Scientists like me thought less": Seaborg and Seaborg, 87.

46   "We were fighting for survival": Ibid.

47   Ten days after Roosevelt approved Bush's plan: Seaborg, *Journals*, 160.

47   "Those who have originated the work on this terrible weapon": Lanouette, 236.

48   "no knowledge at all of nuclear physics": Sanger, 34–35.

48   He told the Met Lab scientists that the construction and operation: Compton, *Atomic Quest*, 109.

CHAPTER 6: PLUTONIUM AT LAST

49   Shortly after coming to Chicago: Seaborg, *Journals*, 131.

49   "We were told to take precautions": Seaborg and Seaborg, 93.

50   "We were pretty young": JaHey, "Those early days as we remember them."

50   Seaborg called it weighing invisible material: Seaborg, *Man-Made Transuranium Elements*, 38.

50   "Today was the most exciting and thrilling day": Seaborg, *Journals*, 176–77.

51   Thomas Edison almost lost his eyesight: Jorgensen, 30–32.

51   By the 1920s, professional societies: Hacker, 13.

52   Seaborg, for example, was always adamant: Seaborg and Seaborg, 75.

52   One time a Met Lab worker was walking by a soda machine: Karle, interview by Alexandra Levy.

53   Once a dog being used in plutonium exposure experiments: Brues, "Those early days as we remember them."

53   Another time the chemists at the Met Lab: Compton, *Memories of Early Atomic Pioneers*, 88.

53   scientists who worked on the Manhattan Project: Voelz et al., 611.

CHAPTER 7: THE DEMONSTRATION

54   Seaborg had a terrifying thought: Seaborg, *Journals*, 201.

55   The army turned up the pressure: Hewlett and Anderson, 106.

56   He said that he could: Seaborg, *Journals*, 216.

57   It was not a squash court: Mort, 1.

57   "This is not it": Allardice and Trapnell, 37.

58   "would probably remember longer than the others": Compton, *Atomic Quest*, 141.

58   Greenewalt later ventured: Greenewalt, interview by Stephane Groueff.

58   "His mind was swarming with ideas": Compton, *Atomic Quest*, 144.

58   "What I was really thinking about": Interview of Crawford Greenewalt by David A. Hounshell and John K. Smith, December 15, 1982, Hagley Museum and Library, Accession 1878. Available upon request to the Hagley Museum and Library.

58   "Of course we have no way of knowing": Seaborg, *Journals*, 217.

60   To avoid perceptions of war profiteering: Hewlett and Anderson, 187.

60   Matthias told his commanding officer: Williams, 17.

PART 2: A FACTORY IN THE DESERT

61   "Some of the flattest, most lonesome territory": Seaborg, *Nuclear Milestones*, 163.

CHAPTER 8: THE EVICTED

63   On March 6, 1943: Gibson, 15.

63   The Wheelers stared: Mendenhall, 407.

64   "It was a wonderful place for the children": Jeanie Shaw Wheeler, inter-

view by Margot Knight, 1979. Available from Manuscripts, Archives, and Special Collections, Washington State University Libraries, Pullman.

64   When the Columbia flooded in spring: "Wheeler Family History," from the collection of family histories in the Public Reading Room, Department of Energy, Richland, WA.

64   In the summers, they could swim: Franklin, 52.

64   "It was a terrible shock," said a resident of Hanford: Sanger, 21.

65   "What's the old barn over there?": Mendenhall, 410.

65   "They appraised my father's 30 acres": Sanger, 21.

65   "If you want to see this place again": Mendenhall, 412.

65   "They came that morning to take us away": Jeanie Shaw Wheeler, interview by Margot Knight, 1979. Available from Manuscripts, Archives, and Special Collections, Washington State University Libraries, Pullman.

66   "I have a promise": Franklin Matthias, interview by B Reactor Museum Association (BRMA), September 26, 1992. Available upon request to BRMA.

66   By the spring of 1941: Sivula, "The man who built Hanford," A6.

67   For the next few years, massive construction equipment: Flynn, 47.

68   On February 26, 1943, he walked into the offices: Williams, 3.

68   "trying to restrict publicity on this project": Findlay and Hevly, 36.

69   On October 16, 1805, Meriwether Lewis: Van Arsdol, *Tri-Cities: The Mid-Columbia Hub*, 10.

69   They called the Columbia *Ci Wana*: Marceau, 1.12.

70   From a population of almost 2,000 in 1780: Ruby and Brown, 11.

70   When they asked for access: Brown, 34.

70   "insist on maintaining their independence": Franklin Matthias diary, April 2, 1944. Available at the Public Reading Room, Department of Energy, Richland, WA.

CHAPTER 9: THE BUILDERS

71   "Suckers, suckers, suckers": Compton, *Memories of Early Atomic Pioneers*, 144.

71   "War Construction Project": Toomey, 60.

72   "Is it true that you people": Groueff, 142.

72   "I had thought there was a lumberjack": Van Arsdol, "Woman recalls strange life in plant's melting-pot camp."

73   It was the largest construction camp: Goldberg, "Groves and the Scientists," 40.

74 The specialized welders known as leadburners: Sanger, 68.

74 By the end of the war, about 15,000 African Americans: Bauman, 124.

75 Hanford also had jobs for women: Gerber, *The Hanford Site*, 11–12.

75 For the men, as one resident said: Sanger, 68.

75 The chief of police noted: Brown, 25.

76 When workers were told: Toomey, 61.

76 "It was exciting": Sasser, interview by Robert Bauman.

76 "This activity, conceived by the workmen": Thayer, 99.

77 "We have a contract with you": Franklin Matthias, interview by B Reactor Museum Association (BRMA), September 26, 1992. Available upon request to BRMA.

77 "Look, take it easy": Thayer, 176.

77 Harry Petcher and his wife Maxine: Sanger, 98–104.

77 "Everything was fine until we got a police call": Ibid., 101.

79 "Second best in class is good": Norris, 49.

79 "If it is a game": Ibid., 41.

80 "Dick has been a wonder of thoughtfulness": Ibid., 58.

80 "Entering West Point fulfilled my greatest ambition": Ibid., 71.

81 Ranked twenty-third in his class: Lawren, 52.

81 "He was really a genius": Franklin Matthias, interview by B Reactor Museum Association (BRMA), September 26, 1992. Available upon request to BRMA.

82 "No one took this man lightly": Norris, 135.

82 He managed the building of army camps and facilities: Bernstein, "Reconsidering the 'Atomic General,'" 894.

83 "I was hoping to get to a war theater": Groves, "Atom General," 16.

83 "The secretary of war has selected you": Groves, *Now It Can Be Told*, 3.

84 That afternoon he argued: Norris, 176.

CHAPTER 10: THE B REACTOR

85 Woods was a child prodigy: Lucibella, 2, 5.

85 "You are a woman, and you will starve to death": Libby, *The Uranium People*, 30.

85 "rather lonesome and empty": Ibid., 84.

86 "Everyone was terrified that . . . the Germans were ahead of us": Libby, interview by S. L. Sanger.

86 "When he told me he was ready": Ibid., 164.

87 The blueprints then were reviewed for accuracy: Ndiaye, 163.

87   "suffered from a general disease": Interview of Crawford Greenewalt by David A. Hounshell and John K. Smith, December 15, 1982, Hagley Museum and Library, Accession 1878. Available upon request to the Hagley Museum and Library.

88   "My rule was simple": Groves, *Now It Can Be Told*, 140.

88   Asking them to "stick to their knitting": Ibid.

88   "You see what I told you?": Groueff, 34.

89   "We were a team, determined to win": Franklin Matthias, from the manuscript "Hanford Engineering Works: Early History," January 14, 1987, p. 11. Available from the Mary Coney papers, Special Collections, University of Washington Libraries, Seattle, WA.

89   "We wore rubber boots": Sanger, 117–18.

90   By May 20, 1944: Hewlett and Anderson, 217.

91   "about as tough as milling iron": B Reactor Museum Association, 27.

91   "They had carloads of Kotex coming in": Ibid., 31.

92   A crucial design decision involved: Johnson et al., 11.

93   Construction workers were cleared: Sanger, 136.

94   "Farmer is his real name": Libby, *The Uranium People*, 163.

94   "With sandwich and apple held high": Ibid., 175.

95   workers in the fuel fabrication plant were producing thousands: Compton, *Memories of Early Atomic Pioneers*, 134.

95   Two men figured out: Grills, interview by Stephane Groueff.

95   The temperature of the metal baths: B Reactor Museum Association, 79.

96   "We worked together to solve problems": Compton, *Memories of Early Atomic Pioneers*, 135.

96   "a drink or two of good whiskey": Libby, *The Uranium People*, 180.

96   Without water running through the tubes: B Reactor Museum Association, 67.

97   "jump right in the middle of it": Lawren, 159.

97   That evening, Fermi and Woods drove back: Libby, *The Uranium People*, 181.

98   "The tale's been told": Compton, *Atomic Quest*, 192–93.

CHAPTER 11: THE T PLANT

101   Leona Woods was entranced: Libby, *The Uranium People*, 185.

102   "an awe-inspiring experience": Seaborg, *Journals*, 463.

102   "One sees nothing but 860 feet": Ibid., 574.

102   The result was about 3,000 gallons of liquid feed: Freer and Conway, 2–4.12.

103   A television camera mounted on the bridge crane: Sanger, 63.

103   "We wanted as few moving parts as possible": Ibid., 61.

103   "That helped," Genereaux said: Ibid., 64.

104   fantastic amounts of gaseous, liquid, and solid waste: Gephart, 5.1–5.54.

104   "great plumes of brown fumes": Libby, *The Uranium People*, 174.

105   the intense neutron flux could make: DeFord, 2–6.14.

106   The initial tanks had lifetimes estimated at 20 years: Gephart, 5.10.

106   "Hesitate, Cogitate, Be Safe": Freer, 2–10.30.

107   "We did a lot of foolish things": Sanger, 190.

108   At four o'clock in the morning on March 2: Hales, 95.

108   "Impossible," he said: Toomey, 64.

108   But as the plans evolved: Harvey and Krafft, 33.

109   He even renamed the planned hotel: Franklin Matthias diary, June 27, 1943. Available at the Public Reading Room, Department of Energy, Richland, WA.

109   "the necessity for maintaining high morale": Pehrson, 42.

109   "Hanford workers were living at the frontier": Findlay and Hevly, 103.

110   Forty-four documented cases: Keating and Harvey, 2–8.10.

CHAPTER 12: IMPLOSION

111   "Do you have a locked compartment": "Something to Win the War: The Hanford Diary," VHS videotape, 1985. Available in the Seattle Room of the Seattle Public Library.

111   "No, I couldn't get a bedroom": Matthias, interview by Stephane Groueff.

113   only about one-seventh of the plutonium atoms split: Aste.

115   In 1944, the spontaneous fission problem: Hoddeson et al., 3.

116   "You don't worry about it": George Kistiakowsky, interview by Richard Rhodes.

116   "Now we have our bomb": Hoddeson et al., 271.

117   In 1944, a Manhattan Project scientist had been startled: Rotblat, interview by Martin Sherwin.

118   "I don't think I'll ever forget D-Day": Crawford Greenewalt. From a speech on the 25th anniversary of the world's first nuclear chain reaction, available in the papers of Herbert L. Anderson, Box 5, Special Collections, University of Chicago Library.

118   In March 1943, he had received a letter: Goudsmit, 47.

119   "The conclusions were unmistakable": Ibid., 70–71.

120   "The capture of this material": Norris, 641, note 69.

120   "The whole German uranium setup": Goudsmit, 106–8.

120   "lacked the fear of an Allied project": Wellerstein, "Historical thoughts."

121   "Isn't it wonderful that the Germans have no atom bomb?": Goudsmit, 76.

CHAPTER 13: WASHINGTON, DC

122   "a discussion was held as to the production schedule": Franklin Matthias diary, March 24, 1945. Available at the Public Reading Room, Department of Energy, Richland, WA.

122   Groves and Matthias had three possible ways: Norris, 368.

122   "Production results have been extremely good": Franklin Matthias diary, May 9, 1945. Available at the Public Reading Room, Department of Energy, Richland, WA.

123   "get the hell out of the road": Sanger, 197.

123   When Eleanor Roosevelt met with Harry Truman: Goodwin, 604.

129   The power of the Interim Committee: Sigal, "Bureaucratic Politics," 329–30.

129   By this time, Groves had been planning: Makhijani, 24.

129   "publicly to affirm its determination": Ellsberg, 226.

130   who later wrote that Szilard had the greatest influence: Seaborg and Seaborg, 118.

134   But the panel member who reportedly held out longest: Wyden, 171.

134   "I believe your people actually *want*": Davis, 182.

135   "We didn't know beans about the military situation in Japan": Bird and Sherwin, 300.

135   The argument that appears to have persuaded: Bernstein, "Roosevelt, Truman and the Atomic Bomb," 61.

135   After the Doolittle raid of April 1942: Prioli, 89.

136   They built thousands of balloons filled with hydrogen: McDowell, 55.

CHAPTER 14: TRINITY

137   extracted from a hot-press mold: Baker et al., 146.

137   They would fit "rather easily": Sanger, 201.

138   director of the test suggested stocking up: Hoddeson et al., 325.

138   A technician accidentally dropped that first initiator: Ibid., 318.

138   He later cited several sources: Rhodes, *Making of the Atomic Bomb*, 572.

139   "important headquarters or troop concentrations": Groves, *Now It Can Be Told*, 267.

141   "It's across the river and it would take a long time to get it": Giovannitti and Freed, 40–41.

141   Opinions differ on exactly why Stimson: Kelly, "Why Did Henry Stimson Spare Kyoto from the Bomb?" 183.

142   "As to the matter of the special target": Norris, 387.

143   On July 24 a message arrived in Washington, DC: Giovannitti and Freed, 247.

144   The endgame had been preordained long before: Bernstein, "Roosevelt, Truman, and the Atomic Bomb," 24.

144   Truman "was like a boy on a toboggan": Knebel and Bailey, 20.

145   He seems to have thought that Hiroshima: Wellerstein, "A 'purely military' target?"

146   On Friday, July 13—a date chosen by George Kistiakowsky: Hoddeson et al., 367.

147   "Let it stick there for a few minutes": Ibid., 369.

147   After one final test of the connections: Kunetka, 300.

148   "I could not absent myself at that time": Compton, *Atomic Quest*, 214.

149   "Groves stayed with the Director": Kunetka, 308.

150   "There was only one dissenting vote": Norris, 662, note 27.

151   a light bright enough to be easily seen from the moon: Reed, 355–56.

151   "at the end of the world:" Hershberg, 234.

151   At the base camp, Fermi was so focused on an experiment: Libby, *The Uranium People*, 229.

152   "The war is over": Groves, *Now It Can Be Told*, 298.

CHAPTER 15: TINIAN ISLAND

153   The previous evening he had watched: Sweeney, 195.

154   "I'd rather face the Japanese than Tibbets": Glines.

155   "It's your call, Chuck": Sweeney, 204–5.

157   Suddenly, as if walking through a doorway: Miller and Spitzer, 97.

157   "We got something wrong here": Bradbury and Blakeslee.

158   "I know all about that": Sweeney, 199.

160   "I can't see the goddamned target": Miller and Spitzer, 101.

160   "No drop," Sweeney said: Sweeney, 213.

160   "Zeros coming up": Ibid., 214.

160   "What's wrong with Nagasaki?": Miller and Spitzer, 103.

161   "What's the score?": Ibid., 106.

161   "It means that, if you agree": Ibid., 108.

161   "I'll guarantee we come within five hundred feet of the target": Sweeney, 217.

161   "Let me think it over, Chuck": Miller and Spitzer, 108.

162   The plane was five minutes from Nagasaki: Coster-Mullen, 76.

162   "I've changed my mind, Chuck": Miller and Spitzer, 109.

162   Jim Van Pelt had his face close: Olivi, 124.

162   "I've got it!": Miller and Spitzer, 110.

162   "You own it": Sweeney, 217.

163   "Never, never again": Miller and Spitzer, 110.

PART 3: UNDER THE MUSHROOM CLOUD

165   "This was the day": Shirabe, *A Physician's Diary*, 1.

CHAPTER 16: NAGASAKI MEDICAL COLLEGE HOSPITAL

167   At exactly eleven o'clock in the morning: From a report written by Raisuke Shirabe in November 1945, translated by Atsumi Nishimura. Available (in Japanese) at https://www-sdc.med.nagasaki-u.ac.jp/abcenter/hibaku -taiken/index.html#taiken1–06.

168   "I cannot describe my thoughts": Shirabe, *A Physician's Diary*, 3.

168   "You're all right, don't worry": Shirabe, "My Experience and Damages." Division of Scientific Data Registry, Atomic Bomb Disease Institute, Nagasaki University. Available (in English) at https://www-sdc.med.nagasaki-u .ac.jp/abcenter/shirabe/index_e.html.

170   "Where were you?": Shirabe, *A Physician's Diary*, 4.

171   One of the students at the medical college: Akizuki, 39.

173   "I sensed a kind of godliness": Shirabe, *A Physician's Diary*, 6.

173   "Two children dragging the body": Yasuyama, 111.

174   He later calculated that the radiation from the bomb: From a report written by Raisuke Shirabe in November 1945, translated by Atsumi Nishimura. Available (in Japanese) at https://www-sdc.med.nagasaki-u .ac.jp/abcenter/hibaku-taiken/index.html#taiken1–06.

174   Anyone standing outside the hospital: Glasstone and Dolan, 333.

174   The bomb's shock wave destroyed: Committee for the Compilation of

Materials on Damage Caused by the Atomic Bombs in Hiroshima and Nagasaki, 40.

176   Houses within five miles of the hypocenter: Tsipis, 87–89.

176   The most recent studies indicate: Robock et al., 9, and Mills et al., 161.

CHAPTER 17: THE URAKAMI VALLEY

177   "Their voices were well harmonized": Shirabe, *A Physician's Diary*, 10.

180   Right after the bombing: Committee for the Compilation of Materials on Damage Caused by the Atomic Bombs in Hiroshima and Nagasaki, 127–85.

181   "Their voices were angry": Shirabe, *A Physician's Diary*, 36.

CHAPTER 18: NAGASAKI

183   They found that within one kilometer: Shirabe et al., *Statistical Observations*, 2.

184   These deaths "may represent death from secondary shock": Ibid., 15.

184   "The atomic bomb is not an inhumane weapon": Southard, 109.

185   In early September, a reporter for the *Chicago Tribune*: Weller, 3–22.

185   "Today the writer spent nearly an hour": Ibid., 31.

185   they were lost for 60 years: Ibid., 306.

185   Fourteen or fifteen people often lived: Southard, 134.

186   "Men and women of the world": Nagai, 118.

PART 4: CONFRONTING ARMAGEDDON

189   "The total elimination of nuclear weapons": Stoffels, "World Citizens for Peace Turns 25."

CHAPTER 19: THE COLD WAR

191   "I couldn't help but believe that God: Van Arsdol, *Hanford: The Big Secret*, 78.

191   even Matthias was initially unclear: Webb, "Local Press Response," 45.

192   "For all we know": Boyer, 5.

192   the *Milwaukee Journal* published a map: Ibid., 14.

193   When a sample of Americans: Boyer, 183.

193   Almost 100,000 Americans died in combat: Clodfelter, 584.

194   Because of the naval blockade: Frank, 149–63.

194   Truman and his aides knew: Kuznick, 9.

194   The entry of the Soviet Union into the war: Hasegawa, 195–201.

194   they initially made little distinction: Wilson, *Five Myths about Nuclear Weapons*, 40–42.

194   Groves's single-minded determination: Goldberg, "Racing to the Finish," 127.

194   Meanwhile, Truman and his future secretary of state: Alperovitz, 270–75.

194   "the most controversial issue in all of American history": Walker, *xi*.

194   The consensus view of most though not all: Ibid., 104.

195   "not to use the weapon on Japan without warning": Seaborg, *Journals*, 700.

195   "I understand and do not quarrel": Seaborg and Seaborg, 120.

196   "I'm proud of you and all your people": Bird and Sherwin, 315.

196   "As the days passed": Smith, *A Peril and a Hope*, 77.

197   "When the second bomb was released": Boyer, 49.

197   "Before God our consciences are clear": Bernstein, "Four Physicists and the Bomb," 241.

197   "We recognized that victory was necessary": Compton, *Atomic Quest*, 277.

197   "As regards criticism of physicists and scientists": Hiltzik, 299.

198   "All [here] are perplexed and bewildered": Fermi, *Atoms in the Family*, 245.

198   "I expect to sleep as well": Boyer, 342.

198   "I have no qualms of conscience": Norris, 426.

198   "I don't think the second bomb": Sanger, 41.

198   For the rest of his life, he wildly exaggerated: Walker, 102.

199   After his death, a researcher discovered: Dower, "Three Narratives," 96.

199   "In the summer of 1945": Rabinowitch, 3.

199   "one of the greatest blunders of history": Lanouette, 276.

202   One anthropologist described America: Boyer, 21.

202   An article in *Life* magazine: "The 36-Hour War," 27–35.

203   "From the river west to Seventh Avenue": Morrison, 8–15.

205   "Discussion of process": Gerber, *On the Home Front*, 47.

205   Military intelligence continued to open mail: Loeb, 26.

206   On September 21, he and his wife: Norris, 444.

207   His final efficiency report said that Groves: Ibid., 504.

208   A 1946 poll found that 54 percent of Americans: Boyer, 37.

208   Within two weeks of Japan's surrender: Herken, 142.

208   "if there are to be atomic weapons in the world": Norris, 473.

209   Early that year, the commissioners: Gerber, *Legend and Legacy*, 18.

CHAPTER 20: BUILDING THE NUCLEAR ARSENAL

210   On September 1, 1949, a B-29: Burr.

210   "Provide us with atomic weapons": Rhodes, *Dark Sun*, 179.

211   According to documents made public: Trenear-Harvey, 133.

211   Using gulag labor and nearby lakes and rivers: Brown, 88–123.

212   Two-thirds of Americans said in a poll: Boyer, 339.

212   "In the brutal and strident climate of the early Cold War": Ibid., 350.

213   "I thought, my lord, what have I gotten myself into?": Ballard, interview by Douglas O'Reagan.

214   A few months later: Ferguson and Smith, 108.

215   They also used the heat from the cooling water: Carlisle and Zenzen, 99.

217   Among the first dramatic groups: Pugnetti, 130.

218   "maternal deaths, infant deaths, and deaths from other causes": Gerber, *Legend and Legacy*, 22.

218   "the Birmingham of Washington": Pihl.

218   in 1950 it had seven Black residents: Findlay and Hevly, 275.

219   less than a dozen African American clerks and custodians: Ibid., 129.

CHAPTER 21: PEAK PRODUCTION

222   When it began generating electricity: Marshall.

222   doubled America's nuclear power capacity: Carlisle and Zenzen, 130.

223   "I assume this is wholly on the level": Cary, "Remembering JFK's visit to Hanford."

223   "Today the wind is at your back": Flenniken, 4.

225   Originally budgeted at $47.7 million: Rhodes, *Energy*, 289.

225   In the 1956 book *Our Friend the Atom*: Cooke, 119.

227   "Fortunately," Seaborg later quipped: Seaborg and Seaborg, 127.

228   "not entirely successful in convincing their teachers": Seaborg, *Man-Made Transuranium Elements*, 17.

228   "with a complete lack of interest": Seaborg and Seaborg, 136.

230   In his final days: Libby, *The Uranium People*, 20.

231   "the most difficult job I ever had": Seaborg and Seaborg, 161.

231   "doubts about the validity": Ibid., 181.

232   On June 7, 1968, Seaborg returned to Richland: Seaborg, *Nuclear Milestones.*

233   Over the previous two years: Seaborg and Seaborg, 240.

234   By 1968, the reactors on the Columbia and Savannah rivers: US Department of Energy, *Plutonium Balance*, 11–13.

236   "We were told to close our eyes": Flenniken, 16.

237   Yet employment at Hanford did not go down much: Findlay and Hevly, 65.

237   *"nobody can lay pipe too slowly"*: Loeb, 117.

CHAPTER 22: THE RECKONING

238   Boosters in the Tri-Cities: Loeb, 110–11.

238   Given estimates made about the same time: Rhodes, *Arsenals of Folly*, 123.

241   In the fall elections: Wittner, 154.

241   "No one expected a big crowd": Godfrey.

244   In 1981, it halted construction: Pope, 163–206.

245   "The accident at Chernobyl": Rhodes, *Arsenals of Folly*, 23.

245   "agree on a stage-by-stage program": Savranskaya and Blanton, "Gorbachev's Nuclear Initiative of January 1986."

246   "It would be fine with me": Rhodes, *Arsenals of Folly*, 265.

248   Since 1943, the federal government had built: Gephart, 5.3–5.16.

249   the waste would deliver a fatal dose: Lloyd et al., 1.

249   plant operators had released: Gephart, 5.25–5.37.

250   In 1986, a group called the Hanford Family: D'Antonio, 200–22.

251   Careful study of Hanford workers: Gilbert et al., "Mortality of Workers," 586–90, and Wing, "Plutonium-related work," 153.

252   studies going all the way back to the Manhattan Project: Voelz et al., 611.

252   nor could studies of populations living nearby: Boice et al., 431.

253   "On the morning I got plucked": Flenniken, 19.

CHAPTER 23: REMEMBERING

255   The department was estimating: Niles, *x*, 60.

256   "no future long-term use": US Department of Energy, Decommissioning of Eight Surplus Production Reactors, 1.1.

256   "We're not in the museum business": Ballard, "70th Anniversary Address."

258   "You have, here in your midst": Rhodes, "Hanford and History."

258   In 2013, unsuccessfully: Smith, *Congressman Doc Hastings*, 369–70.

259   The law firms that the government hired: Greene, 10.

261   "a substantial degree of reassurance to the population": National Academy of Sciences, 121.

261   "These 'reassurances' were worthless": Pritikin, 31.

262   "These findings do not definitively rule out the possibility": Davis et al., 543.

262   Some people undoubtedly got sick and died: Grossman et al., 267.

262   In a hypothetical case where all Americans: National Research Council, 7.

263   Most scientists who study radiation and health: Cravens, 65–79.

263   "This is the greatest honor ever bestowed upon me": LBL Research Review.

264   "Do I wish I hadn't discovered plutonium?": Jolly, 161.

265   more than two-thirds of the 49 people: Argonne National Laboratory, "The Chicago Pile 1 Pioneers."

EPILOGUE

274   One day in 1950: Jones, "Where Is Everybody?," 3.

275   People have proposed dozens: Webb, *If the Universe Is Teeming with Aliens*.

275   Reflecting on the Cuban missile crisis: Allison, "The Cuban Missile Crisis at 50," 11.

275   People may be reassured that the number of warheads: Kristensen and Norris, 289.

275   Even a limited exchange of nuclear weapons: Toon et al., 1224.

275   A full-scale nuclear war would end food production: Robock et al., 9.

276   The reprocessing of reactor fuel to yield plutonium: Feiveson et al., 10–13.

276   All countries will need to agree to a verifiable and enforceable treaty: Daley, 155–88.

278   One time, Fermi was walking: Atomic Heritage Foundation.

BIBLIOGRAPHY

Adams, Melvin R. 2016. *Atomic Geography: A Personal History of the Hanford Nuclear Reservation*. Pullman: Washington State University Press.

Akizuki, Tatsuichiro. 1981. *Nagasaki 1945: The First Full-Length Eyewitness Account of the Atomic Bomb Attack on Nagasaki*. Translated by Keiichi Nagata. New York: Quartet Books.

Allardice, Corbin, and Edward R. Trapnell. 2002. "The First Pile." *Nuclear News* (November): 34–39.

Allison, Graham. 2012. "The Cuban Missile Crisis at 50." *Foreign Affairs* 91 (4): 11–16.

Allison, Samuel K. 1957. "Enrico Fermi, 1901–1954." *Biographic Memoirs*. Washington, DC: National Academy of Sciences.

———. 1965. Interview by Stephane Groueff, Chicago. Voices of the Manhattan Project. Available at https://www.manhattanprojectvoices.org/oral-histories/samuel-k-allisons-interview.

Alperovitz, Gar. 1995. *The Decision to Use the Atomic Bomb and the Architecture of an American Myth*. New York: Knopf.

Alvarez, Luis W. 1987. *Alvarez: Adventures of a Physicist*. New York: Basic Books.

Anderson, Herbert. 1965. Interview by Stephane Groueff. Voices of the Manhattan Project. Available at https://www.manhattanprojectvoices.org/oral-histories/herbert-andersons-interview-1965.

———. 1975. "Assisting Fermi." In *All in Our Time: The Reminiscences of Twelve Nuclear Pioneers*, edited by Jane Wilson, 66–104. Chicago: Bulletin of the Atomic Scientists.

Anderson, Herbert L., as told to J. D. Ratcliff. 1969. "The Day the Atomic Age Was Born." *The Reader's Digest* (March): 129–33.

Argonne National Laboratory. No date. "The Chicago Pile 1 Pioneers." Available at https://www.ne.anl.gov/About/cp1-pioneers/index.html.

Ballard, Del. 2014. "70th Anniversary Address by Del Ballard." Available at https://b-reactor.org.

———. 2016. Interview by Douglas O'Reagan, February 18, Richland. Hanford History Project. Available at http://www.hanfordhistory.com/items/show/183.

Bauman, Robert. 2005. "Jim Crow in the Tri-Cities, 1943–1950." *Pacific Northwest Quarterly* 96 (3): 124–31.

Bauman, Robert, and Robert Franklin, eds. 2018. *Nowhere to Remember: Hanford White Bluffs, and Richland to 1943*. Pullman: Washington State University Press.

Bernstein, Barton. 1975. "Roosevelt, Truman, and the Atomic Bomb, 1941–1945: A Reinterpretation." *Political Science Quarterly* 90 (1): 23–69.

———. 1988. "Four Physicists and the Bomb: The Early Years, 1945–1950." *Historical Studies in the Physical and Biological Sciences* 18 (2): 231–63.

———. 1995. "The Atomic Bombings Reconsidered." *Foreign Affairs* 74 (1): 135–52.

———. 2003. "Reconsidering the 'Atomic General': Leslie R. Groves." *The Journal of Military History* 67 (July): 883–920.

Bernton, Hal. 2017. "Under the mushroom cloud: Will Nagasaki's story be told at Hanford's new national park?" *Seattle Times*, August 3.

Bickel, Lennard. 1979. *The Deadly Element: The Story of Uranium*. New York: Stein and Day.

Bird, Kai, and Lawrence Lifschultz, eds. 1998. *Hiroshima's Shadow: Writings on the Denial of History and the Smithsonian Controversy*. Stony Creek, CT: The Pamphleteer's Press.

Bird, Kai, and Martin J. Sherwin. 2005. *American Prometheus: The Triumph and Tragedy of J. Robert Oppenheimer*. New York: Knopf.

Boice, John D. Jr., Michael T. Mumma, and William J. Blot. 2006. Cancer mortality among populations residing in counties near the Hanford site, 1950–2000. *Health Physics* 90 (5): 431–45.

Boyer, Paul. 1985. *By the Bomb's Early Light: American Thought and Culture at the Dawn of the Atomic Age*. New York: Pantheon.

Bradbury, Ellen, and Sandra Blakeslee. 2015. "The harrowing story of the Nagasaki bombing mission." *The Bulletin of the Atomic Scientists* (August 4).

B Reactor Museum Association. 2001. *Historic American Engineering Record, B Reactor (105-B Building)*. HAER No. WA-164. Richland, WA: US Department of Energy.

Brooks, Antone L. 2018. *Low Dose Radiation: The History of the U.S.*

Department of Energy Research Program. Pullman: Washington State University Press.

Brown, Kate. 2013. *Plutopia: Nuclear Families, Atomic Cities, and the Great Soviet and American Plutonium Disasters.* New York: Oxford University Press.

Brues, Austin M. 1971. "Those early days as we remember them." *Argonne News*, January. Available at https://www.ne.anl.gov/About/early-days/early-days-of-argonne-national-lab.pdf.

Burr, William. 2019. "Detection of the First Soviet Nuclear Test, September 1949." Available at https://nsarchive.gwu.edu/briefing-book/nuclear-vault/2019–09–09/detection-first-soviet-nuclear-test-september-1949.

Bush, Vannevar. 1970. *Pieces of the Action.* New York: William Morrow.

Carlisle, Rodney P., with Joan M. Zenzen. 1996. *Supplying the Nuclear Arsenal: American Production Reactors, 1942–1992.* Baltimore: Johns Hopkins University Press.

Cary, Annette. 2015. "Decades-long Hanford downwinders lawsuit settles." *Tri-City Herald*, October 7.

———. 2017. "Remembering JFK's visit to Hanford. 37,000 watched him wave 'atomic wand.'" *Tri-City Herald*, October 26.

Chinnock, Frank W. 1969. *Nagasaki: The Forgotten Bomb.* New York: World Publishing.

Clodfelter, Michael. 2002. *Warfare and Armed Conflicts: A Statistical Reference to Casualty and Other Figures, 1500–2000.* 2nd ed. London: McFarland.

Collie, Craig. 2011. *Nagasaki: The Massacre of the Innocent and Unknowing.* London: Portobello Books.

Columbia Riverkeeper. 2018. *Competing Visions for Hanford.* Hood River, OR: Columbia Riverkeeper.

Committee for the Compilation of Materials on Damage Caused by the Atomic Bombs in Hiroshima and Nagasaki. 1981. *Hiroshima and Nagasaki: The Physical, Medical, and Social Effects of the Atomic Bombings.* New York: Basic.

Compton, Arthur Holly. 1956. *Atomic Quest: A Personal Narrative.* New York: Oxford University Press.

———. 1967. *The Cosmos of Arthur Holly Compton.* New York: Knopf.

Compton, William M. 2015. *Memories of Early Atomic Pioneers.* Richland, WA: The REACH Museum.

Conant, Jennet. 2005. *109 East Palace: Robert Oppenheimer and the Secret City of Los Alamos.* New York: Simon & Schuster.

Connor, Tim. 1997. *Burdens of Proof: Science and Public Accountability in the Field of Environmental Epidemiology with a Focus on Low Dose Radiation and Community Health Studies*. Energy Research Foundation.

Cooke, Stephanie. 2009. *In Mortal Hands: A Cautionary History of the Nuclear Age*. New York: Bloomsbury.

Copeland, Joe. 2015. *Peace Quest: The Survivors of Hiroshima and Nagasaki*. Smashwords Edition (e-book).

———. 2016. "At Hanford, a chance for a fuller telling of atomic history." *Crosscut*, June 9.

Coster-Mullen, John. 2016. *Atom Bombs: The Top Secret Inside Story of Little Boy and Fat Man*. Self-published.

Cravens, Gwyneth. 2007. *Power to Save the World: The Truth About Nuclear Energy*. New York: Knopf.

Cronin, James W., ed. 2004. *Fermi Remembered*. Chicago: University of Chicago Press.

Cunningham, Jenny. 2018. "A different kind of 'atomic tourist' visits Hanford." *Crosscut*, June 1.

Dahl, Per F. 2002. *From Nuclear Transmutation to Nuclear Fission, 1932–1939*. Philadelphia: Institute of Physics Publishing.

Daley, Tad. 2010. *Apocalypse Never: Forging the Path to a Nuclear Weapon–Free World*. New Brunswick, NJ: Rutgers University Press.

D'Antonio, Michael. 1993. *Atomic Harvest: Hanford and the Lethal Toll of America's Nuclear Arsenal*. New York: Crown.

Davis, Nuel Pharr. 1968. *Lawrence and Oppenheimer*. New York: Simon & Schuster.

Davis, Scott, Kenneth J. Kopecky, Thomas E. Hamilton, Lynn E. Onstad, Beth L. King, Mark S. Saporito, and Christy R. Callahan. 2002. *Hanford Thyroid Disease Study: Final Report*. Seattle: Fred Hutchinson Cancer Research Center.

DeFord, D. H. 2002. "Section 6: Waste Management." In *Hanford Site Historic District: History of the Plutonium Production Facilities, 1943–1990*, 2-6.1–2-6.32. Richland, WA: Battelle Press.

Dower, John W. 1996. "Three Narratives of Our Humanity." In *History Wars: The* Enola Gay *and Other Battles for the American Past*, edited by Edward T. Linenthal and Tom Engelhardt, 63–96. New York: Henry Holt.

DuPont. 1945. *Construction of the Hanford Engineer Works: History of the Project*. HAN-10970, Vol. 4. Wilmington, DE: E. I. Du Pont de Nemours and Company.

Ellsberg, Daniel. 2017. *The Doomsday Machine: Confessions of a Nuclear War Planner*. New York: Bloomsbury.

Feiveson, Harold A., Alexander Glaser, Zia Mian, and Frank N. von Hippel. 2014. *Unmaking the Bomb: A Fissile Material Approach to Nuclear Disarmament and Nonproliferation*. Cambridge, MA: MIT Press.

Feld, Bernard T., and Gertrud Weiss Szilard. 1972. *The Collected Works of Leo Szilard, Scientific Papers*. Cambridge, MA: MIT Press.

Ferguson, Robert L., and C. Mark Smith. 2019. *Something Extraordinary: A Short History of the Manhattan Project, Hanford, and the B Reactor*. Bothell, WA: Book Publishers Network.

Fermi, Enrico. 1965. *Enrico Fermi: Collected Papers, Volume II, United States, 1939–1954*. Chicago: University of Chicago Press.

Findlay, John M., and Bruce Hevly. 2011. *Atomic Frontier Days: Hanford and the American West*. Seattle: University of Washington Press.

Flenniken, Kathleen. 2012. *Plume*. Seattle: University of Washington Press.

Flynn, George. 1993. "The Milwaukee Road's Contribution to the Second World War: The Untold Story of the Milwaukee's Involvement in the Hanford Project." *Milwaukee Railroader* (September): 34–51.

Frank, Richard B. 1999. *Downfall: The End of the Imperial Japanese Empire*. New York: Random House.

Franklin, Robert. 2018. " 'We Worked in the Orchards and We Played in the River': Life in the Towns of Richland, White Bluffs, and Hanford." In *Nowhere to Remember: Hanford White Bluffs, and Richland to 1943*, edited by Robert Bauman and Robert Franklin, 37–62. Pullman: Washington State University Press.

Freer, Brian J. 2002. "Worker Health and Safety." In *Hanford Site Historic District: History of the Plutonium Production Facilities, 1943–1990*, 2-6.1–2-6.32. Richland, WA: Battelle Press.

Freer, Brian J., and Charles A. Conway. 2002. "Chemical Separations." In *Hanford Site Historic District: History of the Plutonium Production Facilities, 1943–1990*, 2-4.1–2-4.32. Richland, WA: Battelle Press.

Frisch, Otto R., and John A. Wheeler. 1967. "The Discovery of Fission." *Physics Today* 20 (11): 43–52.

Gale, Robert Peter, and Eric Lax. 2013. *Radiation: What It Is, What You Need to Know*. New York: Alfred A. Knopf.

Gephart, Roy E. 2003. *Hanford: A Conversation About Nuclear Waste and Cleanup*. Richland, WA: Battelle Press.

Gerber, Michele Stenehjem. 1992. *Legend and Legacy: Fifty Years of Defense Production at the Hanford Site*. Richland, WA: Westinghouse Hanford Company.

———. 1993. *The Hanford Site: An Anthology of Early Histories*. Richland, WA: Westinghouse Hanford Company.

———. 2001. *History of Hanford Site Defense Production.* Richland, WA: Fluor Hanford.

———. 2007. *On the Home Front: The Cold War Legacy of the Hanford Nuclear Site.* Third edition. Lincoln: University of Nebraska Press.

Gibson, Elizabeth. 2002. *Richland, Washington.* Charleston, SC: Arcadia Publishing.

Gilbert, Ethel S., Ellen Omohundro, Jeffrey A. Buchanan, and Nancy A. Holter. 1993. "Mortality of Workers at the Hanford Site: 1945–1986." *Health Physics* 64 (6): 577–90.

Gilbert, Steven, Dianne Dickeman, and Nancy Dickeman. 2011. *Particles on the Wall.* Seattle: Healthy World Press.

Giovannitti, Len, and Fred Freed. 1965. *The Decision to Drop the Bomb.* New York: Coward-McCann.

Glasstone, Samuel, and Philip J. Dolan. 1977. *The Effects of Nuclear Weapons.* Third edition. Washington, DC: United States Department of Defense and United States Department of Energy.

Glines, C. V. 1997. "The bomb that ended the war." *Aviation History Magazine* (January).

Godfrey, Dennis. 1982. "Nearly 100 Attend Richland Rally for Nuclear Arms Freeze." *Tri-City Herald*, August 8.

Goldberg, Stanley. 1992. "Groves Takes the Reins." *Bulletin of the Atomic Scientists* (December): 32–39.

———. 1992. "Inventing a Climate of Opinion: Vannevar Bush and the Decision to Build the Bomb." *Isis* 83(3): 429–52.

———. 1995. "Groves and the Scientists: Compartmentalization and the Building of the Bomb." *Physics Today* 48 (8): 38–43.

———. 1995. "Racing to the Finish: The Decision to Bomb Hiroshima and Nagasaki." *The Journal of American–East Asian Relations* 4 (2): 117–28.

———. 1998. "General Groves and the Atomic West: The Making and Meaning of Hanford." In *The Atomic West*, edited by Bruce Hevly and John M. Findlay, 39–89. Seattle: University of Washington Press.

Goodwin, Doris Kearns. 1994. *No Ordinary Time: Franklin and Eleanor Roosevelt: The Home Front in World War II.* New York: Simon & Schuster.

Gosling, F. G. 2005. *The Manhattan Project: Making the Atomic Bomb.* Washington, DC: U.S. Department of Energy.

Goudsmit, Samuel A. 1947. *Alsos.* New York: Henry Schuman.

Greene, Jenna. 2011. "In Hanford sage, no resolution in sight." *National Law Journal* (June 20).

Greenewalt, Crawford. 1965. Interview by Stephane Groueff, Wilming-

ton, DE. Voices of the Manhattan Project. Available at https://www
.manhattanprojectvoices.org/oral-histories/crawford-greenewalts-interview.

Grills, Raymond. 1965. Interview by Stephane Groueff, Wilmington, January 25. Voices of the Manhattan Project. Available at https://www
.manhattanprojectvoices.org/oral-histories/raymond-grillss-interview.

Grossman, Charles M., Rudi H. Nussbaum, and Fred D. Nussbaum. 2003. "Cancers among Residents Downwind of the Hanford, Washington, Plutonium Production Site." *Archives of Environmental Health* 58 (5): 267–74.

Grossman, Daniel. 1994. "Hanford and Its Early Radioactive Atmospheric Releases." *Pacific Northwest Quarterly* 85 (January):6–14.

Groueff, Stephane. 1967. *Manhattan Project: The Untold Story of the Making of the Atomic Bomb.* Boston: Little, Brown.

Groves, Leslie R. 1948. "The Atom General Answers His Critics." *Saturday Evening Post*, June 19.

———. 1962. *Now It Can Be Told.* New York: Harper.

———. 1965. Interview by Stephane Groueff, January 5. Voices of the Manhattan Project. Available at https://www.manhattanprojectvoices.org/
oral-histories/general-leslie-grovess-interview-part-1.

Guerra, Francesco, Matteo Leone, and Nadia Robotti. 2012. "The Discovery of Artificial Radioactivity." *Physics in Perspective* 14: 33–58.

Guizzo, Erico. 2005. "The Atomic Fortress that Time Forgot." *IEEE Spectrum* 42 (4): 42–49.

Hacker, Barton C. 1987. *The Dragon's Tail: Radiation Safety in the Manhattan Project, 1942–1946.* Berkeley: University of California Press.

Hales, Peter Bacon. 1997. *Atomic Spaces: Living on the Manhattan Project.* Urbana and Chicago: University of Illinois Press.

Ham, Paul. 2014. *Hiroshima, Nagasaki: The Real Story of the Atomic Bombings and Their Aftermath.* New York: St. Martin's.

Hansen, Chuck. 1988. *US Nuclear Weapons: The Secret History.* Arlington, TX: Aerofax.

Hardy, Kenneth. 1974. "Social Origins of American Scientists and Scholars." *Science* 185: 497–506.

Harvey, David. 2000. *History of the Hanford Site: 1943–1990.* Richland, WA: Pacific Northwest National Laboratory.

Harvey, David W., and Katheryn Hill Krafft. 2004. "The Hanford Engineer Works Village: Shaping a Nuclear Community." *Columbia* 18 (1) :29–35.

Hasegawa, Tsuyoshi. 2005. *Racing the Enemy: Stalin, Truman, and the Surrender of Japan.* Cambridge, MA: Harvard University Press.

Hein, Teri. 2000. *Atomic Farmgirl: The Betrayal of Chief Qualchan, the Appaloosa, and Me.* Golden, CO: Fulcrum Publishing.

Herken, Gregg. 2002. *Brotherhood of the Bomb: The Tangled Loyalties of Robert Oppenheimer, Ernest Lawrence, and Edward Teller.* New York: Henry Holt.

Hersey, John. 1946. *Hiroshima.* New York: Alfred A. Knopf.

Hershberg, James G. 1993. *James B. Conant: Harvard to Hiroshima and the Making of the Nuclear Age.* New York: Knopf.

Hevly, Bruce, and John M. Findlay, eds. 1988. *The Atomic West.* Seattle: University of Washington Press.

Hewlett, Richard G., and Oscar E. Anderson, Jr. 1962. *History of the United States Atomic Energy Commission. Volume I. 1939/1946, The New World.* University Park: Pennsylvania State University Press.

Hewlett, Richard G., and Francis Duncan. 1969. *Atomic Shield 1947/1952. Volume II: A History of the United States Atomic Energy Commission.* University Park: Pennsylvania State University Press.

Hiltzik, Michael. 2015. *Big Science: Ernest Lawrence and the Invention That Launched the Military-Industrial Complex.* New York: Simon & Schuster.

Hoddeson, Lillian, Paul W. Henriksen, Roger A. Meade, and Catherine Westfall. 1993. *Critical Assembly: A Technical History of Los Alamos during the Oppenheimer Years, 1943–1945.* New York: Cambridge University Press.

Hoffman, Darleane. 2000. "Glenn Theodore Seaborg." *Biographical Memoirs, Volume 78.* Washington, DC: The National Academies Press.

Hoffman, F. Owen, A. James Ruttenber, A. Iulian Apostoaei, Raymond J. Carroll, and Sander Greenland. 2007. "The Hanford Thyroid Disease Study: An Alternative View of the Findings." *Health Physics* 92 (2): 99–111.

House, Peggy. 1999. "Dr. Seaborg: Citizen-Scholar." *The Seaborg Center Bulletin*, April. Available at https://www.nmu.edu/seaborg/node/9.

JaHey, Arthur H. 1971. "Those early days as we remember them." *Argonne News*, February–March. Available at https://www.ne.anl.gov/About/early-days/early-days-of-argonne-national-lab.pdf.

Johnson, Ben, Richard Romanelli, and Bert Pierard. 2015. *Lost in the Telling: The DuPont Company, The Forgotten Producers of Plutonium.* Richland, WA: B Reactor Museum Association.

Jolly, William L. 1987. *From Retorts to Lasers: The Story of Chemistry at Berkeley.* Berkeley: The College of Chemistry, University of California, Berkeley.

Jones, Eric M. 1985. "'Where Is Everybody?' An Account of Fermi's Question." Los Alamos National Technical Report LA-10311-MS.

Jones, Vincent C. 1985. *Manhattan: The Army and the Atomic Bomb.* Washington, DC: U.S. Government Printing Office.

Jorgensen, Timothy J. 2016. *Strange Glow: The Story of Radiation*. Princeton, NJ: Princeton University Press.

Karle, Isabella. 2015. Interview by Alexandra Levy, Virginia, March 25. Voices of the Manhattan Project. Available at https://www.manhattanprojectvoices.org/oral-histories/isabella-karles-interview-2015.

Kean, Sam. 2019. *The Bastard Brigade: The True Story of the Renegade Scientists and Spies Who Sabotaged the Nazi Atomic Bomb*. New York: Little, Brown.

Keating, J. K., and D. W. Harvey. 2002. "Section 8: Site Security." In *Hanford Site Historic District: History of the Plutonium Production Facilities, 1943–1990*, 2-8.1–2-6.35. Richland, WA: Battelle Press.

Kelly, Cynthia C., ed. 2007. *The Manhattan Project: The Birth of the Atomic Bomb in the Words of Its Creators, Eyewitnesses, and Historians*. New York: Black Dog & Leventhal Publishers.

Kelly, Jason M. 2012. "Why Did Henry Stimson Spare Kyoto from the Bomb?: Confusion in Postwar Historiography." *Journal of American–East Asian Relations* 19 (2): 183–203.

Kiernan, Denise. 2013. *The Girls of Atomic City: The Untold Story of the Women Who Helped Win World War II*. New York: Simon & Schuster.

Kistiakowsky, George. 1982. Interview by Richard Rhodes, Cambridge, January 15. Voices of the Manhattan Project. Available at https://www.manhattanprojectvoices.org/oral-histories/george-kistiakowskys-interview.

Knebel, Fletcher, and Charles W. Bailey. 1963. "The Fight Over the A-Bomb." *Look* (August 13): 19–23.

Krepon, Michael. 2009. *Better Safe Than Sorry: The Ironies of Living with the Bomb*. Stanford, CA: Stanford University Press.

Kristensen, Hans M., and Robert S. Norris. 2017. "Worldwide deployments of nuclear weapons, 2017." *Bulletin of the Atomic Scientists* 73 (5): 289–97.

Kunetka, James. 2015. *The General and the Genius: Groves and Oppenheimer—The Unlikely Partnership that Built the Atom Bomb*. Washington, DC: Regnery.

Kuznick, Peter J. 2007. "The Decision to Risk the Future: Harry Truman, the Atomic Bomb and the Apocalyptic Narrative." *The Asia-Pacific Journal Japan Focus* 5 (7): 1–23.

Lanouette, William, with Bela Silard. 1992. *Genius in the Shadows: A Biography of Leo Szilard, The Man Behind the Bomb*. New York: Charles Scribner's Sons.

Lawren, William. 1988. *The General and the Bomb: A Biography of General Leslie R. Groves, Director of the Manhattan Project*. New York: Dodd, Mead.

LBL Research Review. 1994. "Seaborgium: Element 106 Named in Honor of Glenn T. Seaborg, LBL's Associate Director At Large." August.

Libby, Leona Marshall. 1979. *The Uranium People*. New York: Charles Scribner's Sons.

———. 1986. Interview by S. L. Sanger, Los Angeles. Voices of the Manhattan Project. Available at https://www.manhattanprojectvoices.org/oral-histories/leona-marshall-libbys-interview.

Limerick, Patricia Nelson. "The Significance of Hanford in American History." In *Washington Comes of Age: The State in the National Experience*, edited by David H. Stratton, 153–71. Pullman: Washington State University Press.

Linenthal, Edward T., and Tom Engelhardt, eds. 1996. *History Wars: The Enola Gay and Other Battles for the American Past*. New York: Henry Holt.

Lloyd, W. R., M. K. Sheaffer, and W. G. Sutcliffe. 1994. "Dose Rate Estimates from Irradiated Light-Water-Reactor Fuel Assemblies in Air." Lawrence Livermore National Laboratory.

Loeb, Paul. 1986. *Nuclear Culture: Living and Working in the World's Largest Atomic Complex*. Philadelphia: New Society Publishers.

Lucibella, Michael. 2014. "This Month in Physics History: November 10, 1986: Death of Leona Woods Marshall Libby." *APS News* 23 (10): 2, 5.

Makhijani, Arjun. 1995. " 'Always' the Target: While U.S. Bomb Scientists Were Racing Against Germany, Military Planners Were Looking Toward the Pacific." *Bulletin of the Atomic Scientists* 51 (3): 23–27.

Malloy, Sean L. 2008. *Atomic Tragedy: Henry L. Stimson and the Decision to Use the Bomb Against Japan*. New York: Cornell University Press.

Marceau, T. E. 2002. "Historic Overview." In *Hanford Site Historic District: History of the Plutonium Production Facilities, 1943–1990*, 1.1–1.78. Richland, WA: Battelle Press.

Marshall, Patrick. 2014. "Hanford's N Reactor." Essay 10702. Available at HistoryLink.org.

Marx, Joseph Laurance. 1971. *Nagasaki: The Necessary Bomb?* New York: Macmillan.

Matthias, Franklin. 1965. Interview by Stephane Groueff. Voices of the Manhattan Project. Available at https://www.manhattanprojectvoices.org/oral-histories/colonel-franklin-matthiass-interview-part-1–1965.

Maurer, Ken, Bill Whiting, Marilyn Druby, Kathy Adkisson, and Neal Shulman. *Alive! Yesterday and Today. A History of Richland and the Hanford Project*. Richland, WA.

McDowell, John. 1993. "The Year They Firebombed the West." *American Forests* (May/June): 22–23, 55.

McPike, Kathryn, Tim Malacarne, and Mark Donaldson. 2007. *B Reactor: The First in the World*. Washington, DC: Atomic Heritage Foundation.

Mendenhall, Nancy. 2006. *Orchards of Eden: White Bluffs on the Columbia, 1907–1943*. Seattle: Far Eastern Press.

Miller, Merle, and Abe Spitzer. 1946. *We Dropped the A-Bomb*. New York: Thomas Y. Crowell.

Mills, Michael J., Owen B. Toon, Julia Lee-Taylor, and Alan Robock. 2014. "Multi-decadal global cooling and unprecedented ozone loss following a regional nuclear conflict." *Earth's Future* 2: 161–76.

Miscamble, Wilson D. 2011. *The Most Controversial Decision: Truman, the Atomic Bombs, and the Defeat of Japan*. New York: Cambridge University Press.

Montgomery, Scott L., and Thomas Graham Jr. 2017. *Seeing the Light: The Case for Nuclear Power in the 21st Century*. New York: Cambridge University Press.

Morrison, Philip. 1946. "If the Bomb Gets Out of Hand." In *One World or None: A Report to the Public on the Full Meaning of the Atomic Bomb*, edited by Dexter Masters and Katharine Way, 1–15. New York: McGraw-Hill.

Mort, Howard W. 1957. Memo Pad. *University of Chicago Magazine* 50 (3): 1.

Nagai, Takashi. 1984, originally published in 1949. *The Bells of Nagasaki*. Translated by William Johnston. New York: Kodansha.

Nagasaki Atomic Bomb Testimonial Society. 2016. *Nagasaki: Voices of the A-Bomb Survivors*. Nagasaki: Nagasaki Atomic Bomb Testimonial Society.

Nagasaki University School of Medicine 150th Anniversary Commemoration Group. 2009. *One Hundred and Fifty Years of Modern Medicine in Japan*. Nagasaki: Nagasaki University School of Medicine 150th Anniversary Commemoration Group.

National Academy of Sciences. 2000. *Review of the Hanford Thyroid Disease Study Draft Final Report*. Washington, DC: National Academy Press.

National Park Service and U.S. Department of Energy. 2017. *Foundation Document: Manhattan Project National Historical Park*.

National Research Council. 2006. *Health Risks from Exposure to Low*

Levels of Ionizing Radiation: BEIR VII, Phase 2. Washington, DC: The National Academies Press.

Ndiaye, Pap A. 2006. *Nylon and the Bombs: DuPont and the March of Modern America*. Translated by Elborg Forster. Baltimore: Johns Hopkins University Press.

Nelson, Craig. 2015. *Age of Radiance: The Epic Rise and Dramatic Fall of the Atomic Era*. New York: Scribner.

Niles, Ken. 2014. *Hanford Cleanup: The First 25 Years*. Salem: Oregon Department of Energy.

Nisbet, Jack. 2003. *Visible Bones: Journeys Across Time in the Columbia River Country*. Seattle: Sasquatch Books.

Nobile, Philip, ed. 1995. *Judgment at the Smithsonian: The Bombing of Hiroshima and Nagasaki*. New York: Marlowe.

Norris, Robert S. 2002. *Racing for the Bomb: General Leslie R. Groves, The Manhattan Project's Indispensable Man*. South Royalton, VT: Steerforth Press.

Norris, Robert, and Hans S. Kristensen. 2010. "Global nuclear weapons inventories, 1945–2010." *Bulletin of the Atomic Scientists*, July 1.

Olivi, Fred J. 1999. *Decision at Nagasaki: The Mission That Almost Failed*. Self-published.

Oppenheimer, J. Robert. 1955. *The Open Mind*. New York: Simon & Schuster.

Parides, Peter K. 2009. "Vannevar Bush, James Conant, and the Race to the Bomb—How American Science Was Drafted into Wartime Service." In *The Atomic Bomb and American Society: New Perspectives*, edited by Rosemary B. Mariner and G. Kurt Piehler, 22–41. Knoxville: University of Tennessee Press.

Parker, Martha Berry. 1979. *Tales of Richland, White Bluffs & Hanford 1805–1943 Before the Atomic Reserve*. Fairfield, WA: Ye Galleon Press.

Parker, Herbert M. 1986. *Publications and Other Contributions to Radiological and Health Physics*, edited by Ronald L. Kathren, Raymond W. Baalman, and William J. Bair. Richland, WA: Battelle Press.

Pihl, Kristi. 2011. "Black Tri-Citians reflect on struggles, progress." *Tri-City Herald*, February 14.

Pehrson, G. Albin, and staff. 1943. "Report on the Hanford Engineer Works Village." Available at the Richland Public Library, Richland, WA.

Polmar, Norman, and Robert S. Norris. 2009. *The U.S. Nuclear Arsenal: A History of Weapons and Delivery Systems Since 1945*. Annapolis, MD: Naval Institute Press.

Pope, Daniel. 2008. *Nuclear Implosions: The Rise and Fall of the Wash-*

ington Public Power Supply System. New York: Cambridge University Press.

Power, Max S. 2008. *America's Nuclear Wastelands: Politics, Accountability, and Cleanup.* Pullman: Washington State University Press.

Prioli, Carmine A. 1982. "The Fu-Go Project." *American Heritage Magazine* (April–May): 89–92.

Pritikin, Trisha. 2012. "Insignificant and Invisible: The Human Toll of the Hanford Thyroid Disease Study." In *Tortured Science: Health Studies, Ethics and Nuclear Weapons in the United States,* edited by Dianne Quigley, Amy Lowman, and Steve Wing, 25–52. Amityville, NY: Baywood Publishing.

Pugnetti, Frances Taylor. 1975. *Tiger by the Tail: Twenty-Five Years with the Stormy Tri-City Herald.* Tacoma, WA: Mercury Press.

Rabinowitch, Eugene. 1951. "Five Years After." *Bulletin of the Atomic Scientists,* January.

Reed, Bruce Cameron. 2019. *The History and Science of the Manhattan Project.* Second edition. New York: Springer.

Rhodes, Richard. 1986. *The Making of the Atomic Bomb.* New York: Simon & Schuster.

———. 1995. *Dark Sun: The Making of the Hydrogen Bomb.* New York: Simon & Schuster.

———. 2004. "Hanford and History," keynote address at the B Reactor 60th Anniversary Banquet. Available at http://b-reactor.org/b-reactor-60th-anniversary-address.

———. 2007. *Arsenals of Folly: The Making of the Nuclear Arms Race.* New York: Alfred A. Knopf.

———. 2018. *Energy: A Human History.* New York: Simon & Schuster.

Robock, Alan, Luke Oman, and Georgiy L. Stenchikov. 2007. "Nuclear winter revisited with a modern climate model and current nuclear arsenals: Still catastrophic consequences." *Journal of Geophysical Research: Atmospheres* 112:D13107.

Rotblat, Joseph. 1989. Interview by Martin J. Sherwin, October 16, London. Voices of the Manhattan Project. Available at https://www.manhattanprojectvoices.org/oral-histories/joseph-rotblats-interview.

Ruby, Robert H., and John A. Brown. 1989. *Dreamer-Prophets of the Columbia Basin: Smohalla and Skolaskin.* Norman: University of Oklahoma Press.

Sanger, S. L. 1995. *Working on the Bomb: An Oral History of WWII Hanford.* Portland, OR: Portland State University Continuing Education Press.

Sasser, Norvin. 2013. Interview by Robert Bauman, Northwest Public Television, August 23, Richland, WA. Available at http://www.hanfordhistory.com/items/show/823.

Savranskaya, Svetlana, and Thomas Blanton. 2016. *Gorbachev's Nuclear Initiative of 1986 and the Road to Reykjavik*. National Security Archive Electronic Briefing Book No. 563. Available from https://nsarchive.gwu.edu.

Schwartz, David N. 2017. *The Last Man Who Knew Everything: The Life and Times of Enrico Fermi, Father of the Nuclear Age*. New York: Basic Books.

Schwartz, Stephen I., ed. 1998. *Atomic Audit: The Costs and Consequences of U.S. Nuclear Weapons Since 1940*. Washington, DC: Brookings Institution Press.

Scott, James M. 2015. *Target Tokyo: Jimmy Doolittle and the Raid That Avenged Pearl Harbor*. New York: W. W. Norton.

Seaborg, Glenn T. 1963. *Man-Made Transuranium Elements*. Englewood Cliffs, NJ: Prentice-Hall.

———. 1965. Interview by Stephane Groueff, November 2. Voices of the Manhattan Project. Available at https://www.manhattanprojectvoices.org/oral-histories/glenn-seaborgs-interview.

———. 1972. *Nuclear Milestones: A Collection of Speeches by Glenn T. Seaborg*. San Francisco: W. H. Freeman.

———. 1990. Interview by the American Academy of Achievement, September 21. Available at http://www.achievement.org/achiever/glenn-t-seaborg-ph-d/#interview.

———. 1994. *The Plutonium Story: The Journals of Professor Glenn T. Seaborg, 1939–1946*, edited by Ronald L. Kathren, Jerry B. Gough, and Gary T. Benefiel. Richland, WA: Battelle Press.

———. 1995. "My Career as a Radioisotope Hunter." *JAMA* 273 (12): 961–64.

———. No date. "Glenn T. Seaborg Biography 1912–1999." Available at http://www.seaborg.ucla.edu/biography.html.

Seaborg, Glenn T., with Eric Seaborg. 2001. *Adventures in the Atomic Age: From Watts to Washington*. New York: Farrar, Straus and Giroux.

Segrè, Emilio. 1970. *Enrico Fermi, Physicist*. Chicago: University of Chicago Press.

———. 1983. Interview by Richard Rhodes, June 29, Lafayette, California. Voices of the Manhattan Project. Available at https://www.manhattanprojectvoices.org/oral-histories/emilio-segrès-interview.

Segrè, Gino, and Bettina Hoerlin. 2016. *The Pope of Physics: Enrico Fermi and the Birth of the Atomic Age*. New York: Henry Holt.

Shirabe, Raisuke. 2002. *A Physician's Diary of the Atomic Bombing and Its Aftermath*. Translated by Aloysius F. Kuo. Nagasaki: Nagasaki Association for Hibakusha's Medical Care.

Shirabe, Raisuke, et al. 2006. *Statistical Observations of Atomic Bomb Casualties in Nagasaki*, edited by Yoshisada Shibata. Nagasaki: Nagasaki Association for Hibakusha's Medical Care.

Sigal, Leon V. 1978. "Bureaucratic Politics & Tactical Use of Committees: The Interim Committee & the Decision to Drop the Atomic Bomb." *Polity* 10 (3): 326–64.

———. 1988. *Fighting to a Finish: The Politics of War Termination in the United States and Japan, 1945*. Ithaca, NY: Cornell University Press.

Sime, Ruth Lewin. 2000. "The Search for Transuranium Elements and the Discovery of Nuclear Fission." *Physics in Perspective* 2(1):48–62.

Sivula, Chris. 1992. "Fateful flight selects site." *Tri-City Herald*, December 22, A1.

———. 1993. "The man who built Hanford." *Tri-City Herald*, March 21, A6.

Smith, Alice Kimball. 1970. *A Peril and a Hope: The Scientists' Movement in America: 1945–47*. Cambridge, MA: MIT Press.

Smith, Alice Kimball, and Charles Weiner, eds. 1980. *Robert Oppenheimer: Letters and Recollections*. Cambridge, MA: Harvard University Press.

Smith, Cyril Stanley. 1981. "Some Recollections of Metallurgy at Los Alamos, 1943–45." *Journal of Nuclear Materials* 100: 3–10.

Smith, Mark C. 2018. *Congressman Doc Hastings: Twenty Years of Turmoil*. Bothell, WA: Book Publishers Network.

Smyth, Henry D. 1945. *Atomic Energy for Military Purposes: A General Account of the Development of Methods of Using Atomic Energy for Military Purposes*. Princeton, NJ: Princeton University Press.

Southard, Susan. 2015. *Nagasaki: Life After Nuclear War*. New York: Viking.

Stacy, Ian. 2010. "Roads to Ruin on the Atomic Frontier: Environmental Decision Making at the Hanford Nuclear Reservation, 1942–1952." *Environmental History* 15 (3): 415–48.

Steele, Karen Dorn. 1988. "Hanford's bitter legacy." *Bulletin of the Atomic Scientists* 44 (January/February): 17–23.

Stelson, Caren. 2016. *Sachiko: A Nagasaki Bomb Survivor's Story*. Minneapolis: Carolrhoda Books.

Stoffels, Jim. 2005. "World Citizens for Peace and the Bomb." Available at www.wcpeace.org.

———. 2007. "World Citizens for Peace Turns 25." *Tri-City Herald*, July 29.

Sweeney, Charles W., with James A. Antonucci and Marion K. Antonucci. 1997. *War's End: An Eyewitness Account of America's Last Atomic Mission*. New York: Avon Books.

Szilard, Leo. 1972. *The Collected Works of Leo Szilard*, edited by Bernard T. Feld and Gertrud Weiss Szilard. Cambridge, MA: MIT Press.

———. 1978. *Leo Szilard: His Version of the Facts. Selected Recollections and Correspondence*, edited by Spencer R. Weart and Gertrud Weiss Szilard. Cambridge, MA: MIT Press.

Thayer, Harry. 1996. *Management of the Hanford Engineer Works in World War II: How the Corps, DuPont and the Metallurgical Laboratory Fast Tracked the Original Plutonium Works*. New York: ASCE Press.

"The 36-Hour War." 1945. *Life Magazine* (November 19): 27–35.

Thoennessen, Michael. 2016. *The Discovery of Isotopes: A Complete Compendium*. New York: Springer.

Time Magazine. 1948. "The Eternal Apprentice." November 8, 70–79.

Toomey, Elizabeth. 2015. *The Manhattan Project at Hanford Site*. Charleston, SC: Arcadia Publishing.

Toon, Owen B., Alan Robock, Richard P. Turco, Charles Bardeen, Luke Oman, and Georgiy L. Stenchikov. 2007. "Consequences of Regional-Scale Nuclear Conflicts." *Science* 315 (5816): 1224–25.

Trenear-Harvey, Glenmore S. 2011. *Historical Dictionary of Atomic Espionage*. Lanham, UK: Scarecrow Press.

Tsipis, Kosta. 1983. "Blast, Heat, and Radiation." In *The Nuclear Almanac: Confronting the Atom in War and Peace*, edited by Jack Dennis, 83–97. Reading, MA: Addison-Wesley.

U.S. Department of Energy. 1989. Decommissioning of Eight Surplus Production Reactors at the Hanford Site, Richland, Washington. Draft Environmental Impact Statement.

———. 2012. *The United States Plutonium Balance, 1944–2009*. Washington, DC: US Department of Energy.

———. 2013. *Hanford Site Cleanup Completion Framework*. Richland, WA: US Department of Energy.

Van Arsdol, Ted. 1958. *Hanford: The Big Secret*. Richland, WA: Columbia Basin News.

———. 1990. *Tri-Cities: The Mid-Columbia Hub—An Illustrated History*. Chatsworth, CA: Windsor Publications.

———, Ted. 1993. "Woman recalls strange life in plant's melting-pot camp." *The (Vancouver) Columbian*, March 8.

Van Calmthout, Martijn. 2018. *Sam Goudsmit and the Hunt for Hitler's Atom Bomb*. Amherst, NY: Prometheus.

Voelz, George L., James N. P. Lawrence, and Emily R. Johnson. 1997. "Fifty years of plutonium exposure to the Manhattan Project plutonium workers: An update." *Health Physics* 73 (4): 611–19.

Wahlen, R. K. 1989. *History of the 100-B Area.* Richland, WA: Westinghouse Hanford Company.

Walker, J. Samuel. 2016. *Prompt and Utter Destruction: Truman and the Use of Atomic Bombs against Japan.* Third edition. Chapel Hill: University of North Carolina Press.

Wattenberg, Albert. 1982. "December 2, 1942: the event and the people." *The Bulletin of the Atomic Scientists* (December): 22–32.

Weart, Spencer. 1988. *Nuclear Fear: A History of Images.* Cambridge, MA: Harvard University Press.

Webb, George E. 2009. "Local Press Response to the Atomic Bomb Announcements, August–September 1945." In *The Atomic Bomb and American Society: New Perspectives*, edited by Rosemary B. Mariner and G. Kurt Pichler, 43–64. Knoxville: University of Tennessee Press.

Webb, Stephen. 2015. *If the Universe Is Teeming with Aliens . . . WHERE IS EVERYBODY?: Seventy-Five Solutions to the Fermi Paradox and the Problem of Extraterrestrial Life.* New York: Springer.

Weller, George. 2006. *First into Nagasaki.* New York: Crown.

Wellerstein, Alex. 2015. "The First Light of Trinity." *The New Yorker* (July 16).

———. 2016. "Historical thoughts on Michael Frayn's Copenhagen." *Restricted Data: The Nuclear Secrecy Blog*, February 26, http://blog.nuclearsecrecy.com/2016/02/26/historical-thoughts-michael-frayns-copenhagen.

———. 2018. "A 'purely military' target? Truman's changing language about Hiroshima." *Restricted Data: The Nuclear Secrecy Blog*, January 19, http://blog.nuclearsecrecy.com/2018/01/19/purely-military-target.

Williams, Hill. 2011. *Made in Hanford: The Bomb that Changed the World.* Pullman: Washington State University Press.

Wilson, Jane, ed. 1975. *All in Our Time: The Reminiscences of Twelve Nuclear Pioneers.* Chicago: Bulletin of the Atomic Scientists.

Wilson, Ward. 2013. *Five Myths About Nuclear Weapons.* Boston: Houghton Mifflin Harcourt.

Wing, Steve, and David B. Richardson. 2005. "Age at exposure to ionizing radiation and cancer mortality among Hanford workers: Follow up through 1994." *Occupational and Environmental Medicine* 62: 465–72.

Wing, Steve, David B. Richardson, Susanne H. Wolf, and Gary Mihlan. 2004. "Plutonium-related work and cause-specific mortality at the United

Nagasaki 1945–55, edited by Shunichi Yamashita. Nagasaki: Nagasaki Association for Hibakusha's Medical Care.

Zachary, G. Pascal. 1997. *Endless Frontier: Vannevar Bush, Engineer of the American Century.* New York: Simon & Schuster.

Zorpette, Glenn. 1996. "Hanford's nuclear wasteland." *Scientific American* 274 (5): 88–97.

Zuberi, Matin. 2001. "Atomic Bombing of Hiroshima and Nagasaki." *Strategic Analysis* 25 (5): 623–62.

Zug, James. 2003. *Squash: A History of the Game.* New York: Scribner.

INDEX

Note: Page numbers in *italic* refer to photos, maps, and accompanying captions.